MEMOIRS
of the
American Mathematical Society

Number 1007

Hardy Spaces Associated to Non-Negative Self-Adjoint Operators Satisfying Davies-Gaffney Estimates

Steve Hofmann
Guozhen Lu
Dorina Mitrea
Marius Mitrea
Lixin Yan

November 2011 • Volume 214 • Number 1007 (third of 5 numbers) • ISSN 0065-9266

American Mathematical Society
Providence, Rhode Island

Library of Congress Cataloging-in-Publication Data

Hardy spaces associated to non-negative self-adjoint operators satisfying Davies-Gaffney estimates / Steve Hofmann ... [et al.].

    p. cm. — (Memoirs of the American Mathematical Society, ISSN 0065-9266 ; no. 1007)

    "November 2011, volume 214, number 1007 (third of 5 numbers)."

    Includes bibliographical references.

    ISBN 978-0-8218-5238-5 (alk. paper)

    1. Hardy spaces.   2. Harmonic analysis.   3. Interpolation spaces.   4. Pseudodifferential operators. I. Hofmann, Steve, 1958-

QA320.H37   2011

515.9—dc23   2011030200

Memoirs of the American Mathematical Society

This journal is devoted entirely to research in pure and applied mathematics.

Publisher Item Identifier. The Publisher Item Identifier (PII) appears as a footnote on the Abstract page of each article. This alphanumeric string of characters uniquely identifies each article and can be used for future cataloguing, searching, and electronic retrieval.

Subscription information. Beginning with the January 2010 issue, *Memoirs* is accessible from www.ams.org/journals. The 2011 subscription begins with volume 209 and consists of six mailings, each containing one or more numbers. Subscription prices are as follows: for paper delivery, US\$741 list, US\$592.80 institutional member; for electronic delivery, US\$667 list, US\$533.60 institutional member. Upon request, subscribers to paper delivery of this journal are also entitled to receive electronic delivery. If ordering the paper version, subscribers outside the United States and India must pay a postage surcharge of US\$69; subscribers in India must pay a postage surcharge of US\$95. Expedited delivery to destinations in North America US\$58; elsewhere US\$167. Subscription renewals are subject to late fees. See www.ams.org/help-faq for more journal subscription information. Each number may be ordered separately; *please specify number when ordering an individual number.*

Back number information. For back issues see www.ams.org/bookstore.

Subscriptions and orders should be addressed to the American Mathematical Society, P. O. Box 845904, Boston, MA 02284-5904 USA. *All orders must be accompanied by payment.* Other correspondence should be addressed to 201 Charles Street, Providence, RI 02904-2294 USA.

Copying and reprinting. Individual readers of this publication, and nonprofit libraries acting for them, are permitted to make fair use of the material, such as to copy a chapter for use in teaching or research. Permission is granted to quote brief passages from this publication in reviews, provided the customary acknowledgment of the source is given.

Republication, systematic copying, or multiple reproduction of any material in this publication is permitted only under license from the American Mathematical Society. Requests for such permission should be addressed to the Acquisitions Department, American Mathematical Society, 201 Charles Street, Providence, Rhode Island 02904-2294 USA. Requests can also be made by e-mail to reprint-permission@ams.org.

Memoirs of the American Mathematical Society (ISSN 0065-9266) is published bimonthly (each volume consisting usually of more than one number) by the American Mathematical Society at 201 Charles Street, Providence, RI 02904-2294 USA. Periodicals postage paid at Providence, RI. Postmaster: Send address changes to Memoirs, American Mathematical Society, 201 Charles Street, Providence, RI 02904-2294 USA.

Contents

Abstract

Let X be a metric space with doubling measure, and L be a non-negative, self-adjoint operator satisfying Davies-Gaffney bounds on $L^2(X)$. In this article we present a theory of Hardy and BMO spaces associated to L, including an atomic (or molecular) decomposition, square function characterization, and duality of Hardy and BMO spaces. Further specializing to the case that L is a Schrödinger operator on $\mathbb{R}^n$ with a non-negative, locally integrable potential, we establish additional characterizations of such Hardy spaces in terms of maximal functions. Finally, we define Hardy spaces $H_L^p(X)$ for $p > 1$, which may or may not coincide with the space $L^p(X)$, and show that they interpolate with $H_L^1(X)$ spaces by the complex method.

Received by the editor May 14, 2008 and, in revised form, March 17, 2010.

Article electronically published on March 22, 2011; S 0065-9266(2011)00624-6.

2000 *Mathematics Subject Classification*. Primary 42B20, 42B25; Secondary: 46B70, 47G30.

Key words and phrases. Hardy space, non-negative self-adjoint operator, Davies-Gaffney condition, atom, molecule, BMO, Schrödinger operators, space of homogeneous type.

The authors gratefully acknowledge support from NSF as follows: S. Hofmann (DMS 0245401, DMS 0801079 and FRG 0456306), D. Mitrea (FRG 0456306), M. Mitrea (DMS 0653180 and FRG 0456306), G. Lu (DMS 0500853 and DMS 0901761). L. Yan is supported by an FRG Postdoctoral Fellowship and NSF of China (No. 10771221) and National Natural Science Funds for Distinguished Young Scholar (No. 10925106) .

Affiliations at time of publication: University of Missouri, Department of Mathematics, Columbia, Missouri 65211: Steve Hofmann, email: hofmann@math.missouri.edu; Dorina Mitrea, email: dorina@math.missouri.edu; Marius Mitrea, email: marius@math.missouri.edu.

Guozhen Lu, Department of Mathematics, Wayne State University, Detroit, Michigan 48202; email: gzlu@math.wayne.edu.

Lixin Yan, Department of Mathematics, Zhongshan University, Guangzhou, 510275, People's Republic of China; email: mcsylx@mail.sysu.edu.cn.

CHAPTER 1

Introduction

The development of the theory of Hardy spaces in $\mathbb{R}^n$ was initiated by Stein and Weiss [**SW**], and was originally tied closely to the theory of harmonic functions. On the other hand, real variable methods were introduced into this subject in the seminal paper of Fefferman and Stein [**FS**], the evolution of whose ideas led eventually to a characterization of Hardy spaces via the so called "atomic decomposition", obtained by Coifman [**C**] when $n = 1$, and in higher dimensions by Latter [**L**]. In this context, atoms are compactly supported building blocks, enjoying a vanishing moment condition, whose (countable and suitably convergent) linear combinations generate the entire space. The connection between the results of [**FS**] and those of [**C**] and [**L**] may be seen most directly via the duality pairing with the space BMO, and via the "tent space" theory of Coifman, Meyer and Stein [**CMS**] (in which connection see also the work of Calderón and Torchinsky [**CT**] and Wilson [**Wi**]). The advent of the atomic method enabled the extension of the real variable theory of Hardy spaces to a far more general setting, that of a "space of homogeneous type", in the work of Coifman and Weiss [**CW1**],[**CW2**] (cf. Macias and Segovia [**MS**]). Nonetheless, it is now understood that there are important situations in which the classical Coifman-Weiss theory is not applicable, and these situations, being tied to the theory of partial differential operators generalizing the Laplacian, return us in some sense to the original point of view of [**SW**]. That is, we shall consider Hardy spaces that are adapted to a linear operator L, in much the same way that the classical Stein-Weiss spaces are adapted to the Laplacian. On the other hand, the real variable techniques of [**FS**],[**C**],[**L**], [**CW1**], [**CW2**] and [**CMS**] will still be of fundamental importance to us here.

First Auscher, Duong and McIntosh [**ADM**], and then Duong and Yan, [**DY1**], [**DY2**], introduced Hardy and BMO spaces explicitly adapted to an operator L whose heat kernel enjoys a pointwise Gaussian upper bound (but see also the earlier, more specific work of Auscher and Russ [**AR**]). In their approach, modeled on Duong's earlier work on weak-type $(1, 1)$ bounds for generalized singular integrals (e.g., [**DR**], [**DM**]), the heat semigroup or resolvent replaces the usual averaging operator over cubes or balls (in this connection, see also the work of Martell [**M**] on adapted sharp functions), and, in lieu of a standard vanishing moment condition, "cancellation" becomes a matter of membership in the range of L. Subsequent work on this subject has been based on these two cornerstones.

Recently, in [**AMR**] and in [**HM**], the authors treated Hardy spaces (and in the latter paper, BMO spaces) adapted, respectively, to the Hodge Laplacian on a Riemannian manifold with doubling measure, or to a second order divergence form elliptic operator on $\mathbb{R}^n$ with complex coefficients, in which settings pointwise heat kernel bounds may fail. Thus, although the two cornerstones mentioned above still underlie the foundation of the subject, the results and (to some extent) methods of [**ADM, DY1, DY2**] are not directly applicable. Nonetheless, much of the theory, with some variations, was carried out in [**AMR**] and [**HM**] using only Davies-Gaffney type estimates in place of pointwise kernel bounds. In particular, the adapted H^1 spaces were shown to possess a molecular decomposition, as in the work of Taibleson and Weiss [**TW**]. Molecules are building blocks similar to atoms, but lacking the compact support property of the latter.

In the present work, we extend the results of [**AMR**] in several ways. After treating several preliminary matters in Sections 2 and 3, we develop in Sections 4-6 the theory of H^1 and BMO spaces adapted to an arbitrary non-negative, self-adjoint operator L satisfying Davies-Gaffney bounds, in the general setting of a metric space with a doubling measure, and for our H^1_L space we obtain an atomic decomposition (that is, in which the building blocks are compactly supported). In particular, specializing to the case of the Hodge Laplacian on a Riemannian manifold with doubling measure, this sharpens the result of [**AMR**], who obtain a decomposition in terms of non-compactly supported molecules. To be more precise, we show that the adapted H^1_L spaces defined in terms of atoms, in terms of molecules, or in terms of square functions built with either heat or Poisson semigroups, are all equivalent, assuming sufficient "L-cancellation" of our atoms or molecules. We also establish boundedness of certain maximal operators from our adapted H^1_L space into L^1, although in the absence of any structural assumptions on L, we do not obtain, in this general setting, a maximal function characterization of our space. We then define an adapted BMO_L space, and establish its duality with H^1_L.

We do not address the issue of non-selfadjoint operators as considered in [**HM**]. In the present monograph, self-adjointness is used in two ways: first, to establish an L^2 theory (cf. (3.14) below), and second, to obtain an atomic, as opposed to molecular, decomposition. The first of these is in some sense non-essential: the L^2 theory is available for many non-selfadjoint operators, and were this the only consideration, one could just as well take the L^2 square function bound (3.14) as the fundamental hypothesis, rather than self-adjointness. On the other hand, as regards the second issue, self-adjointness would appear to be essential: we do not necessarily expect that atomic (as opposed to molecular) decompositions will be available in the non-selfadjoint setting. Certainly the method of proof here, based on the wave equation, does not yield such a decomposition without self-adjointness. The atomic decomposition has one particular consequence that we exploit: we also show that an operator T which maps H^1_L molecules uniformly into L^1 is automatically bounded from H^1_L into L^1, without further hypotheses on T; this fact is analogous to results obtained in the classical setting in [**MSV**], [**HZ**], [**HLZ**], [**RV**], [**CYZ**] and [**YZ**], and is non-trivial, in light of the Meyer-Bownik example [**B**]. The proof of this fact uses the atomic decomposition in the following way: we show (cf. Theorem 5.4 below) that a function f given as a finite linear combination of *atoms* has an alternative decomposition as a finite linear combination of *molecules*, with the further property that the ℓ^1 norm of the coefficients in the latter case is actually comparable to the H^1 norm of f. Our other main general result, namely, the equivalence of the molecular H^1_L space and its square function analogues, does not require self-adjointness.

We then proceed to consider certain special cases of the general theory described above. In Section 7, we suppose that the heat kernel enjoys a pointwise Gaussian bound, and prove some sharper results in the spirit of [**DY2**]. Further specializing in Section 8 to the case that L is a Schrödinger operator on $\mathbb{R}^n$, with a non-negative, locally integrable potential, we exploit the explicit structure of the operator to establish additional characterizations of H_L^1 in terms of non-tangential maximal functions built either with heat or Poisson semigroups, by following the methods of [**FS**]. We note that much of the adapted H^1/BMO theory for Schrödinger operators was previously developed in the work of Dziubański and Zienkiewicz [**DZ1, DZ2**] and of Dziubański et al [**DGMTZ**], in the presence of stronger assumptions on the potential. Finally, following [**AMR**], in Section 9 we define the spaces H_L^p, $p > 1$ (which do not necessarily coincide with L^p), and we show that these spaces belong to a complex interpolation scale.

We conclude this introduction by remarking that this work, as well as the earlier cited papers [**DY1**], [**DY2**], [**AMR**], and [**HM**], can in some sense be viewed as a companion to the L^p theory developed for general classes of operators in [**BK1**],[**BK2**], [**HMa**] and [**Au**].

Acknowledgments. The authors thank the referee for a careful reading of the manuscript, and for offering numerous valuable suggestions to improve its mathematical and historical accuracy.

S. Hofmann thanks D. C. Yang for pointing out an error in the original version of Lemma 3.3 of the cited paper [**HM**], which we had quoted without proof (as Lemma 4.3) in an earlier version of this paper. We have addressed this issue here by revising Definitions 2.2 and 2.4, leading now to a correct version of our Lemma 4.3 (Lemma 3.3 of [**HM**]).

L.X. Yan would like to thank X.T. Duong, A. McIntosh and Z. Shen for helpful discussions, and thanks the Department of Mathematics of University of Missouri-Columbia for its hospitality.

CHAPTER 2

Notation and preliminaries

2.1. Spaces of homogeneous type. Throughout the paper we shall make the following standing assumptions:

$$X \text{ is a metric space, with distance function } d, \text{ and}$$
$$(2.1)$$
$$\mu \text{ is a nonnegative, Borel, doubling measure on } X.$$

Recall that a metric is doubling provided that there exists a constant $C > 0$ such that for all $x \in X$ and for all $r > 0$,

$$(2.2) \qquad V(x, 2r) \leq CV(x, r) < \infty,$$

where $B(x, r) := \{y \in X : d(x, y) < r\}$ and

$$(2.3) \qquad V(x, r) := \mu(B(x, r)).$$

In particular, X is a space of homogeneous type. A more general definition[1] and further studies of these spaces can be found in [CW1, Chapter 3]. Note that the doubling property implies the following strong homogeneity property,

$$(2.4) \qquad V(x, \lambda r) \leq C\lambda^n V(x, r)$$

for some $C, n > 0$ uniformly for all $\lambda \geq 1$ and $x \in X$. In Euclidean space with Lebesgue measure, the parameter n corresponds to the dimension of the space, but in our more abstract setting, the optimal n need not even be an integer. There also exist C and D, $0 \leq D \leq n$ so that

$$(2.5) \qquad V(y, r) \leq C\left(1 + \frac{d(x, y)}{r}\right)^D V(x, r)$$

uniformly for all $x, y \in X$ and $r > 0$. Indeed, property (2.5) with $D = n$ is a direct consequence of the triangle inequality for the metric d and the strong homogeneity property (2.4). In the cases of the Euclidean space $\mathbb{R}^n$ and Lie groups of polynomial growth, D can be chosen to be 0.

To simplify notation, we will often just use B for $B(x_B, r_B)$. Also given $\lambda > 0$, we will write λB for the $\lambda-$dilated ball, which is the ball with the same center as B and with radius $r_{\lambda B} = \lambda r_B$. We set

$$(2.6) \qquad U_0(B) := B, \quad \text{and} \quad U_j(B) := 2^j B \backslash 2^{j-1} B \quad \text{for} \quad j = 1, 2, \ldots.$$

For $1 \leq p \leq \infty$, the space of p-integrable functions on X is denoted by $L^p(X)$, the norm of a function $f \in L^p(X)$ by $\|f\|_{L^p(X)}$, and the scalar product in $L^2(X)$ by $\langle ., . \rangle$.

---

[1]We do not treat the more general version of a space of homogeneous type, in which one assumes the existence of a pseudo-metric, rather than a true metric.

2.2. Assumptions. Let (X, d, μ) be as in (2.1). The following will be assumed throughout the paper unless otherwise specified:

(H1) L is a non-negative self-adjoint operator on $L^2(X)$;

(H2) The operator L generates an analytic semigroup $\{e^{-tL}\}_{t>0}$ which satisfies the Davies-Gaffney condition. That is, there exist constants C, $c > 0$ such that for any open subsets U_1, $U_2 \subset X$,

$$(2.7) \qquad |\langle e^{-tL} f_1, f_2 \rangle| \leq C \exp\left(- \frac{\mathrm{dist}(U_1, U_2)^2}{c\,t} \right) \|f_1\|_{L^2(X)} \|f_2\|_{L^2(X)}, \quad \forall\, t > 0,$$

for every $f_i \in L^2(X)$ with $\mathrm{supp}\, f_i \subset U_i$, $i = 1, 2$, where

$$\mathrm{dist}(U_1, U_2) := \inf_{x \in U_1, y \in U_2} d(x, y).$$

2.3. The classical Hardy space $H^1(\mathbb{R}^n)$. It is well-known that the classical Hardy space $H^1(\mathbb{R}^n)$ can be characterized by means of the square or maximal function associated with the Poisson semigroup $e^{-t\sqrt{L}}$ or the heat semigroup e^{-tL}, where $L = -\triangle$ is the Laplace operator, see [**FS**]. A slightly more general point of view is as follows. Let $\psi \in \mathcal{S}(\mathbb{R}^n)$, $\int_{\mathbb{R}^n} \psi \neq 0$, where $\mathcal{S}(\mathbb{R}^n)$ denotes the Schwartz class of smooth functions, rapidly decreasing at infinity. Set $\psi_t(x) := t^{-n} \psi\left(\frac{x}{t}\right)$ for $x \in \mathbb{R}^n$ and $t > 0$. The radial maximal function acting on a tempered distribution $f \in \mathcal{S}'(\mathbb{R}^n)$ is defined as

$$(M_\psi f)(x) := \sup_{t > 0} |(f * \psi_t)(x)|, \quad x \in \mathbb{R}^n.$$

Then f belongs to the Hardy space $H^1(\mathbb{R}^n)$ if and only if $M_\psi f \in L^1(\mathbb{R}^n)$ (see, e.g., [**St2**]).

An important characterization of the Hardy space $H^1(\mathbb{R}^n)$ is in terms of atoms. Recall that a function $a \in L^2(\mathbb{R}^n)$ is called a $H^1(\mathbb{R}^n)$-atom if there exists a ball B in $\mathbb{R}^n$ satisfying

    1) $\mathrm{supp}\, a \subset B$;

    2) $\|a\|_{L^2(\mathbb{R}^n)} \leq |B|^{-1/2}$;

    3) $\int_B a(x)\,dx = 0$.

Here and elsewhere, $|B|$ denotes the Lebesgue measure of the set $B \subset \mathbb{R}^n$. Replacing balls by cubes in (1)-(3) above leads to an equivalent definition.

It is obvious that any $H^1(\mathbb{R}^n)$-atom a is in $H^1(\mathbb{R}^n)$. The basic result about atoms is the following atomic decomposition theorem (see [**C**], [**CW2**] and [**L**]): a real-valued function f defined on $\mathbb{R}^n$ belongs to $H^1(\mathbb{R}^n)$ if and only if it has a decomposition

$$f = \sum_{j=0}^{\infty} \lambda_j a_j \quad \text{in } L^1(\mathbb{R}^n),$$

where the a_j's are $H^1(\mathbb{R}^n)$-atoms and $\sum_{j=0}^{\infty} |\lambda_j| < \infty$. Furthermore,

$$\|f\|_{H^1(\mathbb{R}^n)} \approx \inf \left(\sum_{j=0}^{\infty} |\lambda_j| \right),$$

where the infimum is taken over all such decompositions, and the constants of proportionality are absolute.

2.4. Hardy spaces via atoms. We now introduce the notion of a $(1, 2, M)$-*atom* associated to operators on spaces (X, d, μ) as in (2.1).

DEFINITION 2.1. *Let M be a positive integer. A function $a \in L^2(X)$ is called a $(1, 2, M)$-atom associated to the operator L if there exist a function $b \in \mathcal{D}(L^M)$ and a ball B such that*
(i) $a = L^M b$;
(ii) $\operatorname{supp} L^k b \subset B$, $k = 0, 1, \ldots, M$;
(iii) $\|(r_B^2 L)^k b\|_{L^2(X)} \leq r_B^{2M} V(B)^{-1/2}$, $k = 0, 1, \ldots, M$.

In what follows, let us now assume that
(2.8)
$$M \in \mathbb{N} \quad \text{and} \quad M > \frac{n_0}{4}, \quad \text{where} \quad n_0 := \inf \left\{ n : \sup_{\substack{B \subseteq X \\ \lambda \geq 1}} \left[\frac{V(\lambda B)}{\lambda^n V(B)} \right] < \infty \right\}.$$

I.e., n_0 is the optimal n satisfying (2.4). We denote by $\mathcal{D}(T)$ the domain of an unbounded operator T, and by T^k the k-fold composition of T with itself, in the sense of unbounded operators. Also, let L be as in **(H1)**-**(H2)**.

DEFINITION 2.2. *The atomic Hardy space $H^1_{L,at,M}(X)$ is defined as follows. We shall say that $f = \sum \lambda_j a_j$ is an atomic $(1, 2, M)$-representation (of f) if $\{\lambda_j\}_{j=0}^\infty \in \ell^1$, each a_j is a $(1, 2, M)$-atom, and the sum converges in $L^2(X)$. Set*

$$\mathbb{H}^1_{L,at,M}(X) := \left\{ f : f \text{ has an atomic } (1, 2, M)\text{-representation} \right\},$$

with the norm given by

$$\|f\|_{\mathbb{H}^1_{L,at,M}(X)} =$$
$$\inf \left\{ \sum_{j=0}^\infty |\lambda_j| : f = \sum_{j=0}^\infty \lambda_j a_j \text{ is an atomic } (1, 2, M)\text{-representation} \right\}.$$

The space $H^1_{L,at,M}(X)$ is then defined as the completion of $\mathbb{H}^1_{L,at,M}(X)$ with respect to this norm.

Remark. The assumption of L^2 convergence as a starting point is natural given that we consider here operators for which we assume only an L^2 theory. Indeed, it is not clear that arbitrary ℓ^1 atomic or molecular representations (i.e., for which one does not assume L^2 convergence) make sense in this context. An essentially equivalent, but more complicated method to address this difficulty, in which L^2 convergence of the molecular sums is achieved via truncations of scale, appears in [**HM, HM2**]. An alternative approach, based on convergence of molecular sums in the dual to a BMO-like space, has recently appeared in [**JY**].

2.5. Hardy spaces via molecules. Given (X, d, μ) as in (2.1), M as in (2.8), and $\epsilon > 0$, we next describe the notion of a $(1, 2, M, \epsilon)$-*molecule* associated to an operator L as in **(H1)**-**(H2)**.

DEFINITION 2.3. *A function $m \in L^2(X)$ is called a $(1, 2, M, \epsilon)$-molecule associated to L if there exist a function $b \in \mathcal{D}(L^M)$ and a ball B such that*
(i) $m = L^M b$;
(ii) For every $k = 0, 1, 2, \ldots, M$ and $j = 0, 1, 2, \ldots$, there holds

$$\|(r_B^2 L)^k b\|_{L^2(U_j(B))} \leq r_B^{2M} 2^{-j\epsilon} V(2^j B)^{-1/2},$$

where the annuli $U_j(B)$ have been defined in (2.6).

DEFINITION 2.4. *We fix $\epsilon > 0$. The Hardy space $H^1_{L,mol,M}(X)$ is defined as follows. We say that $f = \sum \lambda_j m_j$ is a molecular $(1, 2, M, \epsilon)$-representation (of f) if $\{\lambda_j\}_{j=0}^\infty \in \ell^1$, each m_j is a $(1, 2, M, \epsilon)$-molecule, and the sum converges in $L^2(X)$. Set*

$$\mathbb{H}^1_{L,mol,M}(X) = \Big\{ f : \ f \text{ has a molecular } (1, 2, M, \epsilon)\text{-representation}\Big\},$$

with the norm given by

$$\|f\|_{\mathbb{H}^1_{L,mol,M}(X)} =$$

$$\inf\Big\{ \sum_{j=0}^\infty |\lambda_j| : f = \sum_{j=0}^\infty \lambda_j m_j \text{ is a molecular } (1, 2, M, \epsilon)\text{-representation}\Big\}.$$

The space $H^1_{L,mol,M}(X)$ is then defined as the completion of $\mathbb{H}^1_{L,mol,M}(X)$ with respect to this norm.

Eventually, we shall see that any fixed choice of $M > n_0/4$ and $\epsilon > 0$, yields the same space.

2.6. Hardy spaces via square and maximal functions. For any $x \in X$ and $\alpha > 0$, the cone of aperture α and vertex x is the set

$$(2.9) \qquad \Gamma^\alpha(x) := \big\{(y, t) \in X \times (0, \infty) : d(y, x) < \alpha t\big\}.$$

For simplicity, we will often write $\Gamma(x)$ in place of $\Gamma^1(x)$. Given an operator L satisfying (**H1**)-(**H2**) and a function $f \in L^1(X)$, consider the following quadratic and non-tangential maximal operators associated to the heat semigroup generated by L

$$(2.10) \qquad S_h f(x) := \Big(\iint_{\Gamma(x)} |t^2 L e^{-t^2 L} f(y)|^2 \frac{d\mu(y)}{V(x,t)} \frac{dt}{t} \Big)^{1/2}, \quad x \in X,$$

and

$$(2.11) \qquad \mathcal{N}_h f(x) := \sup_{(y,t) \in \Gamma(x)} \Big(\frac{1}{V(y,t)} \int_{B(y,t)} |e^{-t^2 L} f(z)|^2 d\mu(z) \Big)^{1/2}, \quad x \in X,$$

where we use an extra averaging in the space variable for the non-tangential maximal function in order to compensate for the lack of pointwise estimates on the heat semigroup (an idea originating in [**KP**]).

One can also consider the Poisson semigroup generated by the operator L and the operators

$$(2.12) \qquad S_P f(x) := \Big(\iint_{\Gamma(x)} |t\sqrt{L} e^{-t\sqrt{L}} f(y)|^2 \frac{d\mu(y)}{V(x,t)} \frac{dt}{t} \Big)^{1/2}, \quad x \in X,$$

and

$$(2.13) \qquad \mathcal{N}_P f(x) := \sup_{(y,t) \in \Gamma(x)} \Big(\frac{1}{V(y,t)} \int_{B(y,t)} |e^{-t\sqrt{L}} f(z)|^2 d\mu(z) \Big)^{1/2}, \quad x \in X,$$

for $f \in L^2(X)$.

In order to define the Hardy spaces based upon these various operators, we follow [**AMR**] and first define the L^2 adapted Hardy space

$$(2.14) \qquad H^2(X) := H_L^2(X) := \overline{R(L)},$$

that is, the closure of the range of L in $L^2(X)$. Then $L^2(X)$ is the orthogonal sum of $H^2(X)$ and the null space $N(L)$. In the sequel, we shall often drop the subscript L when referring to $H^2(X) = H_L^2(X)$.

Before proceeding further, let us observe at this point that there are fairly general circumstances under which $N(L) = \{0\}$, and thus $H^2(X) = L^2(X)$. Indeed, suppose that the space X satisfies the "Ahlfors-David" condition $V(x,t) \approx t^n$, for all $x \in X$ and every $t > 0$ (compare to the weaker (2.4)), and suppose that the heat semigroup e^{-tL} satisfies, for some $p > 2$, the hypercontractive estimate

$$(2.15) \qquad \|e^{-tL}f\|_{L^p(X)} \leq Ct^{\frac{n}{2}(\frac{1}{p}-\frac{1}{2})}\|f\|_{L^2(X)}, \qquad \forall t > 0.$$

Then, writing

$$e^{-tL} - I = \int_0^t \frac{\partial}{\partial s}e^{-sL}ds = -\int_0^t Le^{-sL}ds,$$

we see that $f \in N(L)$ implies that $e^{-tL}f = f$. Consequently, for such f, letting $t \to \infty$ in (2.15), we obtain that $f = 0$, since $f \in L^2(X)$.

We note that, in particular, this last observation shows that if the heat kernel $W_t(x,y)$ of L satisfies the classical pointwise Gaussian bound

$$|W_t(x,y)| \leq Ct^{-n/2}e^{-|x-y|^2/ct},$$

then $H_L^2(X) = L^2(X)$.

Having introduced the space $H^2(X) = H_L^2(X)$, we may now define the spaces $H_{L,S_h}^1(X)$, $H_{L,\mathcal{N}_h}^1(X)$, $H_{L,S_P}^1(X)$, and $H_{L,\mathcal{N}_P}^1(X)$ as the respective completions of $\{f \in H^2(X) : \|Tf\|_{L^1(X)} < \infty\}$, where T denotes, respectively, $S_h, \mathcal{N}_h, S_P$ or $\mathcal{N}_P$, with respect to the norm $\|Tf\|_{L^1(X)}$; e.g.,

$$(2.16) \qquad \|f\|_{H_{L,S_h}^1(X)} := \|S_hf\|_{L^1(X)}, \quad f \in H^2(X),$$

and $H_{L,S_h}^1(X)$ is the completion of $\{f \in H^2(X) : \|S_hf\|_{L^1(X)} < \infty\}$, with respect to the norm defined in (2.16).

Then the following result holds.

THEOREM 2.5. *Suppose $M > \frac{n_0}{4}$ and $\epsilon > 0$. For an operator L satisfying* (**H1**)-(**H2**), *the Hardy spaces $H_{L,S_h}^1(X)$, $H_{L,S_P}^1(X)$, $H_{L,at,M}^1(X)$, and $H_{L,mol,M}^1(X)$ coincide. Furthermore,*

$$\|f\|_{H_{L,S_h}^1(X)} \approx \|f\|_{H_{L,S_P}^1(X)} \approx \|f\|_{H_{L,at,M}^1(X)} \approx \|f\|_{H_{L,mol,M}^1(X)},$$

with implicit constants depending only on n_0, M, ϵ, and L.

2.7. BMO spaces associated to operators. The classical space of functions with bounded mean oscillations on $\mathbb{R}^n$, denoted by $BMO(\mathbb{R}^n)$, was originally introduced by John-Nirenberg in [**JN**]. Recall that a locally integrable function f is said to be in $BMO(\mathbb{R}^n)$ if

$$\|f\|_{BMO(\mathbb{R}^n)} = \sup_B \frac{1}{|B|}\int_B |f(y) - f_B|dy < \infty,$$

where the supremum is taken over all balls $B \subset \mathbb{R}^n$, and f_B stands for the mean of f over B, i.e.

$$f_B = \frac{1}{|B|} \int_B f(y) dy.$$

C. Fefferman and E.M. Stein have proved in [**FS**] that $BMO(\mathbb{R}^n)$ is the dual of $H^1(\mathbb{R}^n)$. For a definition of the BMO space on spaces of homogeneous type we refer the reader to [**CW2**].

Another goal of this paper is to generalize the classical notion of BMO. This generalization is suitably adapted to the operator L and preserves the characteristic properties of the classical BMO spaces, including the duality relationship with the corresponding atomic H^1 space.

In defining our adapted BMO spaces, we follow the approach in [**HM**]. Let $\phi = L^M \nu$ be a function in $L^2(X)$, where $\nu \in \mathcal{D}(L^M)$. For $\epsilon > 0$ and $M \in \mathbb{N}$, we introduce the norm

$$\|\phi\|_{\mathcal{M}_0^{1,2,M,\epsilon}(L)} := \sup_{j \geq 0}\left[2^{j\epsilon} V(x_0, 2^j)^{1/2} \sum_{k=0}^{M} \|L^k \nu\|_{L^2(U_j(B_0))}\right],$$

where B_0 is the ball centered at some $x_0 \in X$ with radius 1, and we set

$$\mathcal{M}_0^{1,2,M,\epsilon}(L) := \{\phi = L^M \nu \in L^2(X) : \|\phi\|_{\mathcal{M}_0^{1,2,M,\epsilon}(L)} < \infty\}.$$

We note that if $\phi \in \mathcal{M}_0^{1,2,M,\epsilon}(L)$ with norm 1, then ϕ is a $(1, 2, M, \epsilon)$-molecule adapted to B_0. Conversely, if m is a $(1, 2, M, \epsilon)$-molecule adapted to any ball, then $m \in \mathcal{M}_0^{1,2,M,\epsilon}(L)$.

Let $(\mathcal{M}_0^{1,2,M,\epsilon}(L))^*$ be the dual of $\mathcal{M}_0^{1,2,M,\epsilon}(L)$, and let A_t denote either $(I + t^2 L)^{-1}$ or $e^{-t^2 L}$. We claim that if $f \in (\mathcal{M}_0^{1,2,M,\epsilon}(L))^*$, then we can define $(I - A_t)^M f$ in the sense of distributions and prove it belongs to $L^2_{\mathrm{loc}}(X)$. Indeed, if $\varphi \in L^2(B)$ for some ball B, it follows from the Davies-Gaffney estimate (2.7) that $(I - A_t)^M \varphi \in \mathcal{M}_0^{1,2,M,\epsilon}(L)$ for every $\epsilon > 0$. Thus,

$$(2.17) \quad \left|\langle (I - A_t)^M f, \varphi \rangle\right| = \left|\langle f, (I - A_t)^M \varphi \rangle\right|$$

$$\leq C_{t, r_B \operatorname{dist}(B, x_0)} \|f\|_{(\mathcal{M}_0^{1,2,M,\epsilon}(L))^*} \|\varphi\|_{L^2(B)} V(B)^{1/2}.$$

Since B was arbitrary, the claim follows. Similarly, $(t^2 L)^M A_t f \in L^2_{\mathrm{loc}}(X)$.

In order to define our adapted BMO spaces we need to introduce one more space. For any $M \in \mathbb{N}$, we set

$$(2.18) \qquad \mathcal{E}_M := \bigcap_{\epsilon > 0} (\mathcal{M}_0^{1,2,M,\epsilon}(L))^*.$$

DEFINITION 2.6. *Suppose $M \geq 1$ and let L be an operator satisfying* (**H1**)-(**H2**). *An element $f \in \mathcal{E}_M$ is said to belong to* $\mathrm{BMO}_{L,M}(X)$ *if*

$$(2.19) \quad \|f\|_{\mathrm{BMO}_{L,M}(X)} := \sup_{B \subset X} \left(\frac{1}{V(B)} \int_B |(I - e^{-r_B^2 L})^M f(x)|^2 d\mu(x)\right)^{1/2} < \infty,$$

where the sup is taken over all balls B in X.

Throughout the paper we make the convention that the space $\mathrm{BMO}_{L,M}(X)$ is understood as classes of functions modulo elements in the null space of the operator L^{M_0}, where M_0 is the least integer strictly bigger than $n_0/4$.

Eventually, we will see that this definition is independent of the choice of $M > n_0/4$ (up to "modding out" elements in the null space of the operator L^{M_0}, as these are annihilated by $(I - e^{-r_B^2 L})^{M_0})$. Compared to the classical definition, in (2.19) the heat semigroup $e^{-r_B^2 L}$ plays the role of averaging over the ball, and the power $M > n_0/4$ provides the necessary cancellation.

The natural analogue of the Fefferman-Stein duality result [**FS**] is the following:

THEOREM 2.7. *Suppose $M \in \mathbb{N}$ and $M > \frac{n_0}{4}$. For an operator L satisfying the conditions* (**H1**)*-*(**H2**)*, there holds (recall the convention made after Definition 2.6)*

$$(H_{L,at,M}^1(X))^* = \mathrm{BMO}_{L,M}(X).$$

The proof of Theorem 2.7 is done in Section 6.

2.8. Historical notes. Hardy and BMO spaces explicitly adapted to an operator L were introduced by Auscher, Duong and McIntosh [**ADM**], and by Duong and Yan, [**DY1**], [**DY2**], in the case that heat kernel of L enjoys a pointwise Gaussian upper bound. The definitions of their adapted Hardy and BMO spaces were similar to those given above, except that the parameter M may always be taken to be 1 in the presence of pointwise kernel bounds. In turn, their approach was modeled on Duong's earlier work on weak-type $(1,1)$ bounds for singular integrals satisfying a generalized Hörmander condition (e.g., [**CD1**], [**DR**], [**DM**]), in which the heat semigroup or resolvent replaces the usual averaging operator over cubes or balls (in this connection, see also the work of Martell [**M**] on adapted sharp functions).

Extensions of the results of [**ADM, DY1, DY2**], to settings in which pointwise kernel bounds may fail, and are replaced by Davies-Gaffney estimates, appear first in [**AMR**] and in [**HM**]. We remark that the present results include those of [**AMR**], in which the Hardy spaces were adapted to a first order Dirac operator D, and were defined in terms of square functions of the form

$$S_\psi f(x) := \left(\iint_{\Gamma(x)} |\psi(tD)f(y)|^2 \frac{d\mu(y)}{V(x,t)} \frac{dt}{t} \right)^{1/2},$$

where ψ has sufficient decay at infinity and sufficient cancellation at the origin. In particular, the choice of $\psi(\zeta) = \zeta^2 e^{-\zeta^2}$ is acceptable, and since $D^2 = \Delta$ (the Hodge Laplacian), one obtains in that case precisely the "heat" square function S_h defined in (2.10), with $L = \Delta$.

CHAPTER 3

Davies-Gaffney estimates

Let (X, d, μ) be as in (2.1). Let $\mathcal{L}(L^p(X), L^q(X))$ stand for the space of bounded linear operators from $L^p(X)$ into $L^q(X)$, for $1 \leq p, q \leq +\infty$, and write $\|T\|_{L^p(X) \to L^q(X)}$ for the operator norm of $T \in \mathcal{L}(L^p(X), L^q(X))$. When $p = q$ we will simply use $\mathcal{L}(L^p(X))$ instead of $\mathcal{L}(L^p(X), L^p(X))$.

3.1. Self-improving properties of Davies-Gaffney estimates. Suppose that, for every $z \in \mathbb{C}_+ = \{z \in \mathbb{C} : \operatorname{Re} z > 0\}$, S_z is a bounded linear operator acting on $L^2(X)$ and that the mapping $\mathbb{C}_+ \ni z \to S_z \in \mathcal{L}(L^2(X))$ is a holomorphic function of z. Assume in addition that

$$(3.1) \qquad \|S_z\|_{L^2(X) \to L^2(X)} \leq 1, \quad \forall z \in \mathbb{C}_+.$$

We say that the family of operators $\{S_z : z \in \mathbb{C}_+\}$ satisfies the Davies-Gaffney estimate if there exist constants C, $c > 0$ such that

$$(3.2) \qquad |\langle S_t f_1, f_2 \rangle| \leq C \exp\left(-\frac{\operatorname{dist}(U_1, U_2)^2}{ct} \right) \|f_1\|_{L^2(X)} \|f_2\|_{L^2(X)}, \quad \forall t > 0,$$

for every $f_i \in L^2(X)$ with $\operatorname{supp} f_i \subset U_i$, $U_i \subset X$, $i = 1, 2$. Of course, the case if $U_1 = U_2 = X$ is just (3.1).

Note that semigroups of operators generated by non-negative self-adjoint operators always satisfy (3.1), and among them many examples of interest satisfy (3.2). Recall that, if L is a non-negative, self-adjoint operator on $L^2(X)$, and $E_L(\lambda)$ denotes its spectral decomposition, then for every bounded Borel function $F : [0, \infty) \to \mathbb{C}$, one defines the operator $F(L) : L^2(X) \to L^2(X)$ by the formula

$$(3.3) \qquad F(L) := \int_0^\infty F(\lambda) dE_L(\lambda).$$

In the case in which $F_z(\lambda) := e^{-z\lambda}$ for $z \in \mathbb{C}_+$, one sets $e^{-zL} := F_z(L)$ as given by (3.3). By the spectral theory, the family $S_z = \{e^{-zL}\}_{z \in \mathbb{C}_+}$ (also called semigroup of operators generated by L) satisfies condition (3.1).

Examples of families of operators for which condition (3.2) holds includes semigroups generated by second order elliptic self-adjoint operators in divergence form, Schrödinger operators with real potential and magnetic field (see, for example [**Si2**]). Condition (3.2) is well-known to hold for Laplace-Beltrami operators on all complete Riemannian manifolds (see [**Da2**],[**Ga**]). In the more general setting of Laplace type operators acting on vector bundles, condition (3.2) is proved in [**Si1**].

Condition (3.2) also holds in the setting of local Dirichlet forms (see, [**Stu**], for instance). In this case the metric measure spaces under consideration are possibly not equipped with any differential structure. However, the semigroups associated

13

with these Dirichlet forms satisfy usually Davies-Gaffney estimates with respect to an intrinsic distance.

PROPOSITION 3.1. *Assume that the operator L satisfies* (**H1**)-(**H2**). *Then for every $K \in \mathbb{N}$, the family of operators*

$$\{(tL)^K e^{-tL}\}_{t>0}$$

satisfies the Davies-Gaffney condition (3.2), with c, $C > 0$ depending on K, n_0 and D only.

We note that, in particular, specializing to the case that $U_1 = U_2 = X$, we have the uniform bound

$$(3.4) \qquad \sup_{t>0} \|(tL)^K e^{-tL}\|_{L^2(X) \to L^2(X)} \le C < \infty.$$

In order to prove Proposition 3.1, we recall a result which appears as Lemma 6.18 in [**Ou**].

LEMMA 3.2. *Suppose that F is an analytic function defined on $\mathbb{C}_+$. Assume that, for two numbers A, $b > 0$,*

$$(3.5) \qquad |F(z)| \le A, \quad \forall z \in \mathbb{C}_+$$

and

$$(3.6) \qquad |F(t)| \le A e^{-\frac{b}{t}}, \quad \forall t > 0.$$

Then for every $z = re^{i\theta}$, $r > 0$ and $\theta \in (-\frac{\pi}{2}, \frac{\pi}{2})$,

$$(3.7) \qquad |F(z)| \le A \exp\left(-\frac{b}{2r}\cos\theta\right).$$

PROOF OF PROPOSITION 3.1. By assumption, L is a non-negative self-adjoint operator on $L^2(X)$. Thus, it follows from spectral theory that the family $S_z = \{e^{-zL}\}_{z \in \mathbb{C}_+}$, the semigroup of operators generated by L, satisfies condition (3.1).

Fix U_1, $U_2 \subset X$ open (not necessarily proper) subsets of X and let f, $g \in L^2(X)$ with $\operatorname{supp} f \subset U_1$ and $\operatorname{supp} g \subset U_2$. Define,

$$F(z) := \langle e^{-zL} f, g \rangle := \int_X e^{-zL} f(x)\overline{g(x)}\, d\mu(x).$$

It follows from the holomorphy of the semigroup on $L^2(X)$ that the function F is holomorphic on $\mathbb{C}_+$. By the Davies-Gaffney condition (3.2),

$$|F(t)| \le e^{-\frac{\operatorname{dist}(U_1,U_2)^2}{ct}} \|f\|_{L^2(U_1)} \|g\|_{L^2(U_2)}, \quad \forall t > 0.$$

In addition, it follows from condition (3.1) that

$$|F(z)| \le \|f\|_{L^2(U_1)} \|g\|_{L^2(U_2)}, \quad \forall z \in \mathbb{C}_+.$$

We then apply Lemma 3.2 to obtain that for every $z = re^{i\theta}$, $r > 0$ and $\theta \in (-\frac{\pi}{2}, \frac{\pi}{2})$,

$$(3.8) \qquad |F(z)| \le \exp\left(-\frac{\operatorname{dist}(U_1,U_2)^2}{2cr}\cos\theta\right) \|f\|_{L^2(U_1)} \|g\|_{L^2(U_2)}, \quad \forall z \in \mathbb{C}_+.$$

The proof of Proposition 3.1 then follows from (3.8) and the Cauchy formula, to the effect that, for every $t > 0$,

$$(tL)^K e^{-tL} = (-1)^K K! \frac{t^K}{2\pi i} \int_{|\zeta - t| = \eta t} e^{-\zeta L} \frac{d\zeta}{(\zeta - t)^{K+1}},$$

where $\eta > 0$ is small enough, and the integral does not depend on η (the choice $\eta = \frac{1}{2}\sin\frac{\theta}{2}$ insures that $\{\zeta : |\zeta - t| \le \eta t\}$ is contained in $\sum(\theta) = \{z \in \mathbb{C} : z \ne 0, |\arg z| < \theta\}$). $\qquad\square$

3.2. Finite speed propagation for the wave equation and Davies-Gaffney estimates. Let L be a non-negative self-adjoint operator. By (3.3) it follows that for every $t > 0$, the operator $\cos(t\sqrt{L})$ is well-defined on $L^2(X)$. Thus it makes sense to make the following definition.

DEFINITION 3.3. *A non-negative self-adjoint operator L is said to satisfy the finite speed propagation property for solutions of the corresponding wave equation if there exists a constant $c_0 > 0$ such that*

$$(3.9) \qquad \langle \cos(t\sqrt{L})f_1,\, f_2 \rangle = 0$$

for all $0 < c_0 t < d(U_1, U_2)$ and $U_i \subset X$, $f_i \in L^2(U_i)$, $i = 1, 2$.

In particular, if $K_{\cos(t\sqrt{L})}(x, y)$ denotes the integral kernel of the operator $\cos(t\sqrt{L})$, then (3.9) entails that for every $t > 0$,

$$(3.10) \qquad \operatorname{supp} K_{\cos(t\sqrt{L})} \subset \mathcal{D}_t := \Big\{ (x, y) \in X \times X : d(x, y) \le c_0 t \Big\}.$$

As a consequence of (3.10), it follows that $K_{\cos(t\sqrt{L})}(x, y) = 0$ for all $(x, y) \notin \mathcal{D}_t$.

PROPOSITION 3.4. *Let L be a non-negative self-adjoint operator acting on $L^2(X)$. Then the finite speed propagation property (3.9) and Davies-Gaffney estimate (3.2) are equivalent.*

PROOF. For the proof, we refer the reader to Theorem 2 in [**Si2**] and Theorem 3.4 in [**CS**]. See also [**CCT**] and [**T**]. $\qquad\square$

Next let L be an operator satisfying (**H1**)-(**H2**). It follows from Proposition 3.4 and (**H2**) that the kernel $K_{\cos(t\sqrt{L})}(x, y)$ of the operator $\cos(t\sqrt{L})$ has the property (3.10). By the Fourier inversion formula, whenever F is an even bounded Borel function with $\hat{F} \in L^1(\mathbb{R})$, we can write $F(\sqrt{L})$ in terms of $\cos(t\sqrt{L})$. Concretely, by recalling (3.3) we have

$$F(\sqrt{L}) = (2\pi)^{-1} \int_{-\infty}^{\infty} \hat{F}(t) \cos(t\sqrt{L})\, dt,$$

which, when combined with (3.10), gives

$$(3.11) \qquad K_{F(\sqrt{L})}(x, y) = (2\pi)^{-1} \int_{|t| \ge c_0^{-1} d(x,y)} \hat{F}(t) K_{\cos(t\sqrt{L})}(x, y)\, dt.$$

LEMMA 3.5. *Let $\varphi \in C_0^{\infty}(\mathbb{R})$ be even, $\operatorname{supp}\varphi \subset (-c_0^{-1}, c_0^{-1})$, where c_0 is the constant in (3.10). Let Φ denote the Fourier transform of φ. Then for every $\kappa = 0, 1, 2, \ldots$, and for every $t > 0$, the kernel $K_{(t^2 L)^\kappa \Phi(t\sqrt{L})}(x, y)$ of $(t^2 L)^\kappa \Phi(t\sqrt{L})$ satisfies*

$$(3.12) \qquad \operatorname{supp} K_{(t^2 L)^\kappa \Phi(t\sqrt{L})}(x, y) \subseteq \Big\{ (x, y) \in X \times X : d(x, y) \le t \Big\}.$$

PROOF. For every $\kappa = 0, 1, 2, \ldots$, we set $\Psi_{\kappa,t}(\zeta) := (t\zeta)^{2\kappa}\Phi(t\zeta)$. Using the definition of the Fourier transform, it can be verified that

$$\widehat{\Psi_{\kappa,t}}(s) = (-1)^\kappa \frac{1}{t}\psi_\kappa\Big(\frac{s}{t}\Big),$$

where we have set $\psi_\kappa(s) = \frac{d^{2\kappa}}{ds^{2\kappa}}\varphi(s)$. Observe that for every $\kappa = 0, 1, 2, \ldots$, the function $\Psi_{\kappa,t} \in \mathcal{S}(\mathbb{R})$ is an even function. It follows from formula (3.11) that

$$(3.13) \quad K_{(t^2 L)^\kappa \Phi(t\sqrt{L})}(x, y) = (-1)^\kappa \frac{1}{2\pi} \int_{|st| \geq c_0^{-1} d(x,y)} \frac{d^{2\kappa}}{ds^{2\kappa}}\varphi(s) K_{\cos(st\sqrt{L})}(x, y)\, ds.$$

Since $\varphi \in C_0^\infty(\mathbb{R})$ and $\operatorname{supp} \varphi \subset (-c_0^{-1}, c_0^{-1})$, the claim in Lemma 3.5 follows readily from this. $\qquad\square$

Finally, for $s > 0$, we define

$$\mathbb{F}(s) := \left\{ \psi : \mathbb{C} \to \mathbb{C} \text{ measurable} : \quad |\psi(z)| \leq C \frac{|z|^s}{(1 + |z|^{2s})} \right\}.$$

Then for any non-zero function $\psi \in \mathbb{F}(s)$, we have that $\{\int_0^\infty |\psi(t)|^2 \frac{dt}{t}\}^{1/2} < \infty$. Denote by $\psi_t(z) = \psi(tz)$. It follows from the spectral theory in [**Yo**] that for any $f \in L^2(X)$,

$$\left\{ \int_0^\infty \|\psi(t\sqrt{L})f\|_{L^2(X)}^2 \frac{dt}{t} \right\}^{1/2} = \left\{ \int_0^\infty \langle \overline{\psi}(t\sqrt{L})\, \psi(t\sqrt{L})f, f \rangle \frac{dt}{t} \right\}^{1/2}$$

$$= \left\{ \langle \int_0^\infty |\psi|^2(t\sqrt{L}) \frac{dt}{t} f, f \rangle \right\}^{1/2}$$

$$(3.14) \qquad\qquad \leq \kappa \|f\|_{L^2(X)},$$

(with equality if $f \in H^2(X)$) where $\kappa = \{\int_0^\infty |\psi(t)|^2 dt/t\}^{1/2}$, an estimate which will be used often in the sequel.

CHAPTER 4

The decomposition into atoms

The aim of this chapter is to show that the "square function" and "atomic" H^1 spaces are equivalent, if the parameter $M > n_0/4$. In fact, we shall prove

THEOREM 4.1. *Suppose that $M > n_0/4$. Then $H^1_{L,at,M}(X) = H^1_{L,S_h}(X)$. Moreover,*

$$\|f\|_{H^1_{L,at,M}} \approx \|f\|_{H^1_{L,S_h}},$$

where the implicit constants depend only on M, n_0 and on the constants in the Gaffney and doubling conditions.

Consequently, one may write $H^1_{L,at}$ in place of $H^1_{L,at,M}$, when $M > n_0/4$, as these spaces are all equivalent. In fact, more generally, given Theorem 4.1, we have the following:

DEFINITION 4.2. *The Hardy space $H^1_L(X)$ is the space*

$$H^1_L(X) := H^1_{L,S_h}(X) = H^1_{L,at}(X) := H^1_{L,at,M}(X), \quad M > n_0/4.$$

4.1. Strategy of the proof of Theorem 4.1.

OUTLINE OF THE PROOF. Recall that $H^1_{L,at,M}(X)$ and $H^1_{L,S_h}(X)$ are, respectively, the completions of $\mathbb{H}^1_{L,at,M}(X)$ and of $H^1_{L,S_h}(X) \cap H^2(X)$.

We proceed in two stages: first, to show that $\mathbb{H}^1_{L,at,M}(X) \subset (H^2(X) \cap H^1_{L,S_h}(X))$, with

$$\|f\|_{H^1_{L,S_h}(X)} \leq C\|f\|_{\mathbb{H}^1_{L,at,M}(X)}$$

(this is the content of Proposition 4.4 below); and second, to show the opposite containment with the reverse inequality (this is the content of Proposition 4.13 below). Thus, the two completions $H^1_{L,at,M}(X)$ and $H^1_{L,S_h}(X)$ have the same dense subset $\mathbb{H}^1_{L,at,M}(X) = H^1_{L,S_h}(X) \cap H^2(X)$, with equivalence of norms, and are therefore the same space. The details of this two stage argument follow (respectively) in the next two subsections. $\square$

Before proceeding to the proof of Theorem 4.1, we record now for future reference two observations.

First, we note that the operator S_h is bounded on $L^2(X)$. Indeed, for every $f \in L^2(X)$,

$$\int_X |S_h f(x)|^2 d\mu(x) = \int_X \int_0^\infty \int_{d(x,y)<t} |t^2 L e^{-t^2 L} f(y)|^2 \frac{d\mu(y)}{V(x,t)} \frac{dt}{t} d\mu(x)$$

$$(4.1) \qquad \approx \int_0^\infty \int_X |t^2 L e^{-t^2 L} f(y)|^2 \frac{d\mu(y) dt}{t} \leq C\|f\|_{L^2(X)}^2,$$

where the last step in (4.1) is a particular case of (3.14), and the next-to-last step is obtained by using condition (2.5) to deduce that, for $d(x, y) < t$,

$$(4.2) \qquad \int_{d(x,y)<t} V(x,t)^{-1} d\mu(x) \approx \int_{d(x,y)<t} V(y,t)^{-1} d\mu(x) = 1.$$

Next, we note the following technical lemma.

LEMMA 4.3. *Fix $M \in \mathbb{N}$. Assume that T is a linear operator, or a non-negative sublinear operator, satisfying the weak-type $(2,2)$ bound*

$$(4.3) \qquad \mu\{x \in X : |Tf(x)| > \eta\} \le C_T \, \eta^{-2} \|f\|_{L^2(X)}^2, \qquad \forall \eta > 0,$$

and that for every $(1, 2, M)$-atom a, we have

$$(4.4) \qquad \|Ta\|_{L^1(X)} \le C$$

with constant C independent of a. Then T is bounded from $\mathbb{H}^1_{L,at,M}(X)$ to $L^1(X)$, and

$$\|Tf\|_{L^1(X)} \le C \|f\|_{\mathbb{H}^1_{L,at,M}(X)}.$$

Consequently, by density, T extends to a bounded operator from $H^1_{L,at,M}(X)$ to $L^1(X)$.

PROOF. Let $f \in \mathbb{H}^1_{L,at,M}(X)$, where $f = \sum \lambda_j a_j$ is an atomic $(1, 2, M)$-representation such that

$$\|f\|_{\mathbb{H}^1_{L,at,M}(X)} \approx \sum_{j=0}^{\infty} |\lambda_j|.$$

Since the sum converges in L^2 (by Definition 2.2), and since T is of weak type $(2, 2)$, we have that at almost every point,

$$(4.5) \qquad |T(f)| \le \sum_{j=0}^{\infty} |\lambda_j| \, |T(a_j)|,$$

with equality without absolute value if T is linear. Indeed, for every $\eta > 0$, we have that, if $f^N := \sum_{j>N} \lambda_j a_j$, then,

$$\mu\Big\{ |T(f)| - \sum_{j=0}^{\infty} |\lambda_j| \, |T(a_j)| > \eta \Big\} \le \limsup_{N \to \infty} \mu\left\{ |T(f^N)| > \eta \right\}$$

$$\le C_T \, \eta^{-2} \limsup_{N \to \infty} \|f^N\|_2^2 = 0,$$

from which (4.5) follows. In turn, (4.5) and (4.4) imply the desired L^1 bound for Tf. The last claim in the statement is a routine consequence of the non-negative sublinearity of T and the triangle inequality, much as in the case when T is linear. $\qquad \square$

We now are ready to present the two stage proof of Theorem 4.1

4.2. $\mathbb{H}^1_{L,at,M}(X) \subseteq H^1_{L,S_h}(X) \cap H^2(X)$ **for all** $M > n_0/4$. This stage of the proof is contained in the following Proposition.

PROPOSITION 4.4. *Suppose that* $M > \frac{n_0}{4}$ *and that* L *satisfies* **(H1)**-**(H2)**. *Then* $\mathbb{H}^1_{L,at,M}(X) \subset H^1_{L,S_h}(X) \cap H^2(X)$, *and*

$$\|f\|_{H^1_{L,S_h}(X)} \leq C\|f\|_{\mathbb{H}^1_{L,at,M}(X)}$$

for some $C = C(M, n_0) > 0$.

PROOF. We begin by noting that $\mathbb{H}^1_{L,at,M}(X) \subset H^2(X)$. Indeed, by definition, a $(1, 2, M)$-atom belongs to $R(L)$ (in fact, to $R(L^M)$), and therefore so does any finite linear combination of atoms. Moreover, by definition, every $f \in \mathbb{H}^1_{L,at,M}(X)$ is an L^2 limit of such a finite linear combination, whereby $f \in \overline{R(L)} = H^2(X)$.

It remains to show that the square function maps $\mathbb{H}^1_{L,at,M}(X)$ into L^1. To this end, we observe that by Lemma 4.3, it will be enough to show that for every $(1, 2, M)$-atom a associated to a ball B of X, we have $\|S_h a\|_{L^1(X)} \leq C$. By Hölder's inequality, we may write

$$(4.6) \qquad \|S_h a\|_{L^1(X)} \leq C \sum_{j=0}^{\infty} V(2^j B)^{1/2} \|S_h a\|_{L^2(U_j(B))}.$$

Since S_h is bounded on $L^2(X)$, we can write

$$(4.7) \qquad \|S_h a\|_{L^2(U_j(B))} \leq C\|a\|_{L^2(B)} \leq CV(B)^{-1/2}, \quad \text{for } j = 0, 1, 2.$$

Fix some $j \geq 3$. We note that since a is a $(1, 2, M)$-atom associated to the ball B, by definition, there exists a function $b \in \mathcal{D}(L^M)$, such that $a = L^M b$, which satisfies (ii) and (iii) in Definition 2.1. We then estimate the L^2 norm of $S_h a$ on $U_j(B)$ by decomposing the domain of integration as follows.

$$\|S_h a\|^2_{L^2(U_j(B))}$$

$$= \int_{U_j(B)} \int_0^\infty \int_{d(x,y)<t} |t^2 L e^{-t^2 L} a(y)|^2 \frac{d\mu(y)}{V(x,t)} \frac{dt}{t} d\mu(x)$$

$$= \int_{U_j(B)} \int_0^\infty \int_{d(x,y)<t} |(t^2 L)^{M+1} e^{-t^2 L} b(y)|^2 \frac{d\mu(y)}{V(x,t)} \frac{dt}{t^{4M+1}} d\mu(x)$$

$$= \int_{U_j(B)} \left(\int_0^{r_B} + \int_{r_B}^{d(x,x_B)/4} + \int_{d(x,x_B)/4}^\infty \right) \int_{d(x,y)<t} \dots d\mu(y) \, dt \, d\mu(x)$$

$$=: I_j + II_j + III_j, \quad \text{respectively.}$$

Let us first estimate the term I_j. Set

$$(4.8) \qquad E_j(B) := \{y \in X : d(x,y) \leq r_B \text{ for some } x \in U_j(B)\}.$$

If $z \in B$ and $y \in E_j(B)$, then for $x \in U_j(B)$ with $d(x,y) \leq r_B$ we have

$$d(y,z) \geq d(x,x_B) - d(x,y) - d(z,x_B)$$

$$\geq d(x,x_B) - 2r_B \geq d(x,x_B)/2 \geq 2^{j-2} r_B,$$

and, thus, $\mathrm{dist}(E_j(B), B) \geq 2^{j-2} r_B$. Using estimate (4.2) and Proposition 3.1 with $K = M + 1$, $\|b\|_{L^2(B)} \leq r_B^{2M} V(B)^{-1/2}$ and (2.8), we see that

$$
\begin{aligned}
I_j &\leq C \int_0^{r_B} \int_{E_j(B)} \left|(t^2 L)^{M+1} e^{-t^2 L} b(y)\right|^2 d\mu(y) \frac{dt}{t^{4M+1}} \\
&= C \int_0^{r_B} \left\|(t^2 L)^{M+1} e^{-t^2 L} b\right\|_{L^2(E_j(B))}^2 \frac{dt}{t^{4M+1}} \\
&\leq C \|b\|_{L^2(B)}^2 \int_0^{r_B} \exp\left(-\frac{\mathrm{dist}(E_j(B), B)^2}{ct^2}\right) \frac{dt}{t^{4M+1}} \\
&\leq C r_B^{4M} V(B)^{-1} \int_0^{r_B} \left(\frac{t}{2^j r_B}\right)^{4M+1} \frac{dt}{t^{4M+1}} \\
&\leq C 2^{-j(4M+1-n_0)} V(2^j B)^{-1} \left[2^{-jn_0} \frac{V(2^j B)}{V(B)}\right] \\
&\leq C 2^{-j(4M-n_0)/2} V(2^j B)^{-1},
\end{aligned}
$$

which is of the right order. In order to estimate the second term II_j, observe that if $z \in B$ and

$$
y \in F_j(B) := \left\{y \in X : d(x, y) \leq \frac{d(x, x_B)}{4} \text{ for some } x \in U_j(B)\right\},
$$

then for $x \in U_j(B)$ with $d(x, y) \leq \frac{d(x, x_B)}{4}$ we have

$$
d(y, z) \geq d(x, x_B) - d(x, y) - d(z, x_B)
$$
$$
\geq \frac{3d(x, x_B)}{4} - r_B \geq \frac{d(x, x_B)}{2} \geq 2^{j-3} r_B,
$$

and hence $\mathrm{dist}(F_j(B), B) \geq 2^{j-3} r_B$. Estimate (4.2), together with Proposition 3.1 and the condition $M > n_0/4$ shows that

$$
\begin{aligned}
II_j &\leq C \int_{r_B}^\infty \int_{F_j(B)} \left|(t^2 L)^{M+1} e^{-t^2 L} b(y)\right|^2 d\mu(y) \frac{dt}{t^{4M+1}} \\
&\leq C \|b\|_{L^2(B)}^2 \int_{r_B}^\infty \exp\left(-\frac{\mathrm{dist}(F_j(B), B)^2}{ct^2}\right) \frac{dt}{t^{4M+1}} \\
&\leq C r_B^{4M} V(B)^{-1} \int_{r_B}^\infty \left(\frac{t}{2^j r_B}\right)^{2M+\frac{n_0}{2}} \frac{dt}{t^{4M+1}} \\
&\leq C 2^{-j(4M-n_0)/2} V(2^j B)^{-1},
\end{aligned}
$$

which suits our purpose. Finally, for the term III_j we obtain

$$
\begin{aligned}
III_j &\leq C \int_{2^{j-1} r_B}^\infty \int_X \left|(t^2 L)^{M+1} e^{-t^2 L} b(y)\right|^2 d\mu(y) \frac{dt}{t^{4M+1}} \\
&\leq C \int_{2^{j-1} r_B}^\infty \frac{dt}{t^{4M+1}} \|b\|_{L^2(B)}^2 \\
&\leq C 2^{-4Mj} V(B)^{-1} \leq C 2^{-j(4M-n_0)/2} V(2^j B)^{-1}
\end{aligned}
$$

by the condition $M > n_0/4$. Combining the estimates for I_j, II_j and III_j obtained above, we may conclude that for every $j \geq 3$,

$$
\|S_h a\|_{L^2(U_j(B))} \leq C 2^{-j(4M-n_0)/4} V(2^j B)^{-1/2}.
$$

The latter, together with (4.6), (4.7), and the condition $M > n_0/4$, gives that $\|S_h a\|_{L^1(X)} \le C$. We have therefore proved that $a \in H^1_{L,S_h}(X)$ with $\|a\|_{H^1_{L,S_h}(X)} \le C$. Hence, the proof of Proposition 4.4 is completed. $\qquad\square$

REMARK 4.5. *It turns out that, assuming Gaussian upper estimates for the heat kernel of the operator L, we can take $M = 1$ in Proposition 4.4 and in other similar results (this observation was made previously in* [**AMR**]*). We will come back to this point in Section 7.*

We now turn to the reverse estimate.

4.3. The inclusion $(H^1_{L,S_h}(X) \cap H^2(X)) \subseteq \mathbb{H}^1_{L,at,M}(X)$ **for all** $M \ge 1$. The aim of this section is to establish an atomic $(1, 2, M)$-representation for functions in the space $H^1_{L,S_h}(X) \cap H^2(X)$. This atomic decomposition will be obtained using Lemma 3.5 and adapting the arguments in [**CMS**] and [**Ru**] to the present situation.

4.3.1. *Tent spaces on spaces of homogeneous type.* We begin by reviewing tent spaces on X following [**CMS**] and [**Ru**]. For any $x \in X$ and $\alpha > 0$, recall (2.9) and for any closed subset $F \subseteq X$ define a saw-tooth region $\mathcal{R}^\alpha(F) := \bigcup_{x \in F} \Gamma^\alpha(x)$. For simplicity we will write $\mathcal{R}(F)$ instead of $\mathcal{R}^1(F)$. If O is an open subset of X, and we denote by E^c the complement of a set E, then the "tent" over O, denoted by $\widehat{O}$, is defined as

$$(4.9) \qquad \widehat{O} := \left[\mathcal{R}(O^c)\right]^c = \{(x, t) \in X \times (0, \infty) : d(x, O^c) \ge t\}.$$

LEMMA 4.6. *For a measurable function F defined on $X \times (0, \infty)$, consider*

$$(4.10) \qquad \mathcal{A}^\alpha F(x) := \left(\iint_{\Gamma^\alpha(x)} |F(y, t)|^2 \frac{d\mu(y)}{V(x, \alpha t)} \frac{dt}{t} \right)^{1/2}, \qquad \alpha > 0,$$

and set $\mathcal{A}F(x) = \mathcal{A}^1 F(x)$. Then there exists a constant $C > 0$ depending only on n_0 in (2.8) and the constant D in (2.5) of X such that

$$\|\mathcal{A}^\alpha F\|_{L^1(X)} \le C\|\mathcal{A}F\|_{L^1(X)}.$$

PROOF. The argument is similar to that of Proposition 4 in [**CMS**] corresponding to the case $X = \mathbb{R}^n$. See also Theorem 7.1 in [**FOS**]. $\qquad\square$

Following [**CMS**] and [**Ru**], given $0 < p < \infty$, the "tent space" $T^p_2(X)$ is defined as the space of measurable functions F on $X \times (0, \infty)$, for which $\mathcal{A}F \in L^p(X)$. This is equipped with $\|F\|_{T^p_2(X)} := \|\mathcal{A}F\|_{L^p(X)}$. Observe that $T^p_2(X)$ is a Banach space when $p \in [1, \infty)$.

For future reference, we note that for any compact set K in $X \times (0, \infty)$, and for $1 \le p < \infty$, we have

$$(4.11) \qquad \int_K |F(x, t)|^2 d\mu(x)\, dt \le C(K, p)\|\mathcal{A}(F)\|^2_{L^p(X)},$$

as one may verify, in the more delicate case $p < 2$, by observing that the doubling property implies that

$$|F(x, t)| \approx \int_{d(x,y)<t} |F(x, t)| V(y, t)^{-1} d\mu(y),$$

and then using Minkowski's integral inequality and compactness of K.

The duality for tent spaces is as follows:

PROPOSITION 4.7. *The pairing $\langle F, G \rangle \mapsto \int_{X \times (0,\infty)} F(x,t)G(x,t)d\mu(x)dt/t$ realizes $T_2^{p'}(X)$ as equivalent with the dual of $T_2^p(X)$ if $1 < p < \infty$ and $1/p + 1/p' = 1$.*

In the sequel, $[\,,\,]_\theta$ and $(\,,\,)_{\theta,q}$ denote the complex and real method of interpolation described in [**BL**], respectively. Then we have the following results of interpolation of tent spaces.

PROPOSITION 4.8. *Suppose $1 \le p_0 < p < p_1 \le \infty$, with $1/p = (1-\theta)/p_0 + \theta/p_1$ and $0 < \theta < 1$. Then*

$$\left[T_2^{p_0}(X), T_2^{p_1}(X)\right]_\theta = T_2^p(X)$$

$$and \quad \left(T_2^{p_0}(X), T_2^{p_1}(X)\right)_{\theta,q} = T_2^p(X), \; if\, p = q.$$

Next we review the atomic theory for tent spaces as originally developed in [**CMS**], and extended to the setting of spaces of homogeneous type in [**Ru**].

DEFINITION 4.9. *A measurable function A on $X \times (0,\infty)$ is said to be a T_2^1-atom if there exists a ball $B \subset X$ such that A is supported in $\widehat{B}$ (defined in (4.9)) and*

$$(4.12) \qquad \iint_{X \times (0,\infty)} |A(x,t)|^2 d\mu(x) \frac{dt}{t} \le \frac{1}{V(B)}.$$

Note that if A is a T_2^1-atom supported in $\widehat{B}$, then we have

$$(4.13) \quad \|A\|_{T_2^1(X)} = \int_X \left(\iint_{\Gamma^\alpha(x)} |A(y,t)|^2 \frac{d\mu(y)\,dt}{V(x,t)t} \right)^{1/2} d\mu(x)$$

$$= \int_{C_\alpha B} \left(\iint_{\Gamma^\alpha(x)} |A(y,t)|^2 \frac{d\mu(y)\,dt}{V(x,t)t} \right)^{1/2} d\mu(x)$$

$$\le C_\alpha V(B)^{1/2} \left(\int_X \iint_{d(x,y)<\alpha t} |A(y,t)|^2 \frac{d\mu(y)\,dt}{V(x,t)t} d\mu(x) \right)^{1/2}$$

$$\le C_\alpha V(B)^{1/2} \left(\iint_{X \times (0,\infty)} \left(\int_{d(x,y)<\alpha t} \frac{1}{V(x,t)} d\mu(x) \right) |A(y,t)|^2 \frac{d\mu(y)\,dt}{t} \right)^{1/2}$$

$$\le C_\alpha,$$

where for the last inequality in (4.13) we have used (4.2) and (4.12).

It has been proved in [**Ru**] that every $F \in T_2^1(X)$ has an atomic decomposition. For future reference, we record this result below.

PROPOSITION 4.10. *For every element $F \in T_2^1(X)$ there exist a numerical sequence $\{\lambda_j\}_{j=0}^\infty \subset \ell^1$ and a sequence of T_2^1-atoms $\{A_j\}_{j=0}^\infty$ such that*

$$(4.14) \qquad F = \sum_{j=0}^\infty \lambda_j A_j \quad in\ T_2^1(X)\ and\ a.e.\ in\ X \times (0,\infty).$$

Moreover,

$$\sum_{j=0}^\infty |\lambda_j| \approx \|F\|_{T_2^1(X)},$$

where the implicit constants depend only on the homogeneous space properties of X.

Finally, if $F \in T_2^1(X) \cap T_2^2(X)$, then the decomposition (4.14) also converges in $T_2^2(X)$.

PROOF. Except for the final part of the proposition, concerning T_2^2 convergence, this is Theorem 1.1 in [**Ru**], and we refer the reader to that paper for the proof. The T_2^2 convergence is only implicit there, so we shall sketch the proof here. To this end, we first note that by (4.2) (cf. (4.1)), we have

$$(4.15) \quad \|F\|_{T_2^2(X)}^2 := \int_X (\mathcal{A}F)^2 d\mu = \int_X \int_0^\infty \int_{d(x,y)<t} |F(y,t)|^2 \frac{d\mu(y)}{V(x,t)} \frac{dt}{t} d\mu(x)$$

$$\approx \int_0^\infty \int_X |F(y,t)|^2 \frac{d\mu(y)dt}{t}$$

Suppose now that $F \in T_2^1 \cap T_2^2$. We recall that, in the constructive proof of the decomposition (4.14) in [**Ru**], one has that

$$\lambda_j A_j = F 1_{S_j},$$

where $\{S_j\}$ is a collection of sets which are pairwise disjoint (up to sets of measure zero), and whose union covers $X \times (0,\infty)$. Thus, by (4.15),

$$\Big\| \sum_{j>N} \lambda_j A_j \Big\|_{T_2^2(X)}^2 \approx \int_0^\infty \int_X \Big| \sum_{j>N} 1_{S_j} F(y,t) \Big|^2 \frac{d\mu(y)dt}{t}$$

$$= \sum_{j>N} \iint_{S_j} |F|^2 \frac{d\mu(y)dt}{t} \to 0,$$

as $N \to \infty$, where we have used disjointness of the sets S_j and dominated convergence. It therefore follows that $F = \sum \lambda_j A_j$ in T_2^2. $\square$

Next, we discuss some preliminary matters en route to the atomic decomposition of $H_{L,S_h}^1(X) \cap H^2(X)$ (Proposition 4.13 below). Let $M \geq 1$, and for the remainder of this section let φ, c_0, and Φ be as in Lemma 3.5, but with the added assumptions that $\varphi \geq 0$, with $\varphi \geq c > 0$ on $(-1/(2c_0), 1/(2c_0))$. Set $\Psi(x) := x^{2(M+1)}\Phi(x)$, $x \in \mathbb{R}$. Consider the operator $\pi_{\Psi,L} : T_2^2(X) \to L^2(X)$, given by

$$(4.16) \qquad \pi_{\Psi,L}(F)(x) := \int_0^\infty \Psi(t\sqrt{L})\big(F(\cdot,t)\big)(x)\frac{dt}{t},$$

where the improper integral converges weakly in L^2. The bound

$$(4.17) \qquad \|\pi_{\Psi,L}F\|_{L^2(X)} \leq C_M \|F\|_{T_2^2(X)}, \quad M \geq 0,$$

follows readily by duality and the L^2 quadratic estimate (3.14).

Moreover, we have the following analogue of the well-known argument of Theorem 6 of [**CMS**].

LEMMA 4.11. *Suppose that A is a $T_2^1(X)$-atom associated to a ball $B \subset X$ (or more precisely, to its tent $\widehat{B}$). Then for every $M \geq 1$, there is a uniform constant C_M such that $C_M^{-1} \pi_{\Psi,L}(A)$ is a $(1,2,M)$-atom associated to the concentric double $2B$.*

PROOF. Fix a ball B and let A be a $T_2^1(X)$-atom A associated to $\widehat{B}$. Thus,

$$\int_{X \times (0,\infty)} |A(x,t)|^2 d\mu(x)dt/t \leq V(B)^{-1}.$$

We write

$$a := \pi_{\Psi,L}(A) = L^M b,$$

where

$$b := \int_0^\infty t^{2M} t^2 L\Phi(-t\sqrt{L})\big(A(\cdot, t)\big)\frac{dt}{t}.$$

We observe that the functions $L^k b$, $k = 0, 1, ..., M$, are supported on the ball $2B$, by Lemma 3.5, since A is supported in $\widehat{B}$. Consider some $g \in L^2(2B)$ such that $\|g\|_{L^2(2B)} = 1$. Then for every $k = 0, 1, \ldots, M$ we have

$$(4.18) \quad \left| \int_X (r_B^2 L)^k b(x)\, g(x) d\mu(x) \right|$$

$$= \left| \lim_{\delta \to 0} \int_X \left(\int_\delta^{1/\delta} t^{2M} r_B^{2k} L^k t^2 L\Phi(-t\sqrt{L})\big(A(\cdot, t)\big)(x)\frac{dt}{t} \right) g(x)\, d\mu(x) \right|$$

$$= \left| \int_{\widehat{B}} A(x,t) t^{2M} r_B^{2k} L^k t^2 L\Phi(-t\sqrt{L}) g(x)\frac{d\mu(x)dt}{t} \right|$$

$$\leq r_B^{2M} \|A\|_{T_2^2(X)} \left(\int_{\widehat{B}} \left| (t^2 L)^{k+1}\Phi(-t\sqrt{L}) g(x) \right|^2 \frac{d\mu(x)dt}{t} \right)^{1/2}$$

$$\leq C r_B^{2M} V(B)^{-1/2} \|g\|_{L^2(2B)}.$$

Here, the third line is obtained by using the compactness of the t interval to interchange the order of integration, and the fourth line by using that A is a T_2^1-atom supported in $\widehat{B}$ (hence, $0 < t < r_B$), and that $k \leq M$. The last inequality follows from (3.14). We then have the $(1, 2, M)$-atomic bounds

$$\|(r_B^2 L)^k b\|_{L^2(2B)} \quad \leq \quad C r_B^{2M} V(B)^{-1/2}, \quad k = 0, 1, ..., M,$$

finishing the proof. $\qquad\square$

We will also use the following elementary fact.

LEMMA 4.12. *Let B_1, B_2 be Banach spaces, and let T be a bounded linear operator from B_1 into B_2. Suppose that $\sum F_i$ converges in B_1. Then $\sum f_i := \sum T(F_i)$ converges in B_2.*

PROOF. We have that

$$\limsup_{N \to \infty} \left\| \sum_{i=0}^N f_i - T\left(\sum_{i=0}^\infty F_i\right) \right\|_{B_2}$$

$$= \limsup_{N \to \infty} \left\| T\left(\sum_{i=N+1}^\infty F_i\right) \right\|_{B_2} \leq C \limsup_{N \to \infty} \left\| \sum_{i=N+1}^\infty F_i \right\|_{B_1} = 0,$$

so that the desired conclusion follows. $\qquad\square$

We are now ready to establish the atomic decomposition of $H^1_{L,S_h}(X) \cap H^2(X)$. Our argument here follows the (now standard) tent space approach of [**CMS**], as modified in [**AMR**] to treat the situation in which pointwise kernel bounds may be lacking[1]. A similar approach, again following [**CMS**] and [**AMR**], appears in [**JY**]. A more complicated adaptation of the methods of [**CMS**] was used in [**HM, HM2**].

PROPOSITION 4.13. *Suppose $M \geq 1$ and L satisfies* (**H1**)-(**H2**). *If $f \in H^1_{L,S_h}(X) \cap H^2(X)$, then there exist a family of $(1,2,M)$-atoms $\{a_j\}_{j=0}^{\infty}$ and a sequence of numbers $\{\lambda_j\}_{j=0}^{\infty} \subset \ell^1$ such that f can be represented in the form $f = \sum_{j=0}^{\infty} \lambda_j a_j$, with the sum converging in $L^2(X)$, and*

$$\|f\|_{\mathbb{H}^1_{L,at,M}(X)} \leq C \sum_{j=0}^{\infty} |\lambda_j| \leq C\|f\|_{H^1_{L,S_h}(X)},$$

where C is independent of f. In particular,

(4.19)
$$H^1_{L,S_h}(X) \cap H^2(X) \subseteq \mathbb{H}^1_{L,at,M}(X).$$

PROOF. Let $f \in H^1_{L,S_h}(X) \cap H^2(X)$, and set

$$F(\cdot,t) := t^2 L e^{-t^2 L} f.$$

We note that $F \in T^2_2(X) \cap T^1_2(X)$, by (3.14) and the definition of $H^1_{L,S_h}(X)$. Therefore, by Proposition 4.10, we have that

$$F = \sum \lambda_j A_j,$$

where each A_j is a T^1_2-atom, the sum converges in both $T^1_2(X)$ and $T^2_2(X)$, and

(4.20)
$$\sum |\lambda_j| \leq C\|F\|_{T^1_2(X)} = C\|f\|_{H^1_{L,S_h}(X)}.$$

Also, by L^2-functional calculus ([**Mc**]), and using that $f \in H^2(X)$, we have the "Calderón reproducing formula"

(4.21) $$f(x) = c_\Psi \int_0^\infty \Psi(t\sqrt{L})(t^2 L e^{-t^2 L} f)(x) \frac{dt}{t}$$

$$= c_\Psi\, \pi_{\Psi,L}(F) = c_\Psi \sum \lambda_j\, \pi_{\Psi,L}(A_j),$$

where by (4.17) and Lemma 4.12 the last sum converges in $L^2(X)$. Moreover, by Lemma 4.11, for every $M \geq 1$, we have that up to multiplication by some harmless constant C_M, each $a_j := c_\Psi\, \pi_{\Psi,L}(A_j)$ is a $(1,2,M)$-atom. Consequently, the last sum in (4.21) is an atomic $(1,2,M)$-representation, so that $f \in \mathbb{H}^1_{L,at,M}(X)$, and by (4.20) we have

$$\|f\|_{\mathbb{H}^1_{L,at,M}(X)} \leq C\|f\|_{H^1_{L,S_h}(X)}.$$

The proof is completed. $\qquad\square$

4.4. Equivalence of $H^1_{L,S_P}(X)$ and $H^1_{L,at,M}(X)$ when $M > n_0/4$. We recall that $H^1_{L,S_P}(X)$ is defined to be the completion of the set $\{f \in H^2(X) : \|S_P f\|_{L^1(X)} < \infty\}$, with respect to the norm

$$\|f\|_{H^1_{L,S_P}(X)} := \|S_P f\|_{L^1(X)},$$

where the operator S_P is defined in (2.12). We have the following:

THEOREM 4.14. *Suppose that $M > n_0/4$. Then $H^1_{L,S_P}(X) = H^1_{L,S_h}(X) = H^1_{L,at,M}(X)$, with equivalence of norms.*

---

[1]In particular, it is the idea of [**AMR**] to exploit the fact that a T^1_2-atomic decomposition, of an element in $T^1_2 \cap T^2_2$, converges also in T^2_2.

We start with the following auxiliary result.

LEMMA 4.15. *Fix a number $K \in \mathbb{N}$. For all closed sets E, F in X with* $\mathrm{dist}(E, F) > 0$, *and with $f \in L^2(X)$ supported in E, we have*

$$\left\| (t\sqrt{L})^{2K} e^{-t\sqrt{L}} \right\|_{L^2(F)} \le C \left(\frac{t}{\mathrm{dist}(E, F)} \right)^{2K+1} \| f \|_{L^2(E)}, \quad \forall t > 0,$$

and also

$$\left\| (t\sqrt{L})^{2K+1} e^{-t\sqrt{L}} \right\|_{L^2(F)} \le C \left(\frac{t}{\mathrm{dist}(E, F)} \right)^{2K+1} \| f \|_{L^2(E)}, \quad \forall t > 0.$$

PROOF. The proof in [**HM**] of a similar result deals with the case of divergence form elliptic operators in $\mathbb{R}^n$, but carries out to the present context mutatis mutandis. We therefore give only a brief sketch of the argument.

The subordination formula

$$(4.22) \qquad e^{-t\sqrt{L}} f = \frac{1}{\sqrt{\pi}} \int_0^\infty \frac{e^{-u}}{\sqrt{u}} e^{-\frac{t^2}{4u} L} f \, du$$

allows us to estimate

$$\left\| (t\sqrt{L})^{2K} e^{-t\sqrt{L}} \right\|_{L^2(F)} \le C \int_0^\infty \frac{e^{-u}}{\sqrt{u}} \left\| \left(\frac{t^2 L}{4u} \right)^K e^{-\frac{t^2}{4u} L} f \right\|_{L^2(F)} u^K \, du$$

$$(4.23) \qquad\qquad\qquad \le C \| f \|_{L^2(E)} \int_0^\infty e^{-u} e^{-\frac{u\, \mathrm{dist}(E,F)^2}{ct^2}} u^{K-1/2} \, du.$$

Then we make the change of variables $u \mapsto s := u \frac{\mathrm{dist}(E,F)^2}{ct^2}$ to bound the last term in (4.23) by

$$C \| f \|_{L^2(E)} \int_0^\infty e^{-s \frac{t^2}{\mathrm{dist}(E,F)^2}} e^{-s} \left(s \frac{t^2}{\mathrm{dist}(E,F)^2} \right)^{K-1/2} \frac{t^2}{\mathrm{dist}(E,F)^2} \, ds$$

$$\le C \| f \|_{L^2(E)} \left(\frac{t}{\mathrm{dist}(E,F)} \right)^{2K+1} \int_0^\infty e^{-s} s^{K-1/2} \, ds$$

$$\le C \left(\frac{t}{\mathrm{dist}(E,F)} \right)^{2K+1} \| f \|_{L^2(E)}$$

which proves the first estimate in the conclusion of the lemma. To prove the second estimate, we note that $t\sqrt{L} e^{-t\sqrt{L}} f = -t\, \partial(e^{-t\sqrt{L}} f)/\partial t$, so that the subordination formula now yields

$$t\sqrt{L} e^{-t\sqrt{L}} f = 2 \frac{1}{\sqrt{\pi}} \int_0^\infty \frac{e^{-u}}{\sqrt{u}} \frac{t^2 L}{4u} e^{-\frac{t^2}{4u} L} f \, du.$$

The rest of the argument follows as before. $\qquad\square$

We now turn to the Hardy spaces $H^1_{L,S_P}(X)$ defined in terms of the square function associated to the Poisson semigroup $\{ e^{-t\sqrt{L}} \}_{t>0}$. As was the case for the space $H^1_{L,S_h}(X)$, it is enough to establish the following analogue of Propositions 4.4 and 4.13.

PROPOSITION 4.16. *Let L be an operator satisfying* (**H1**)-(**H2**), *and suppose that $M > n_0/4$. Then $H^1_{L,S_P}(X) \cap H^2(X) = \mathbb{H}^1_{L,at,M}(X)$, with equivalence of norms. More precisely, we have*

(i) Suppose $M > \frac{n_0}{4}$. Then $\mathbb{H}^1_{L,at,M}(X) \subseteq H^1_{L,S_P}(X) \cap H^2(X)$, and

$$\|f\|_{H^1_{L,S_P}(X)} \leq C\|f\|_{\mathbb{H}^1_{L,at,M}(X)},$$

for some C independent of f.

(ii) Conversely, if $f \in H^1_{L,S_P}(X) \cap H^2(X)$, then for every $M \geq 1$, there exist a family of $(1,2,M)$-atoms $\{a_j\}_{j=0}^{\infty}$ and a sequence of numbers $\{\lambda_j\}_{j=0}^{\infty} \subset \ell^1$ such that f can be represented in the form $f = \sum_{j=0}^{\infty} \lambda_j a_j$, with the sum converging in $L^2(X)$, and

$$\|f\|_{\mathbb{H}^1_{L,at,M}(X)} \leq C \sum_{j=0}^{\infty} |\lambda_j| \leq C\|f\|_{H^1_{L,S_P}(X)},$$

where C is independent of f. In particular,

$$(4.24) \qquad H^1_{L,S_P}(X) \cap H^2(X) \subseteq \mathbb{H}^1_{L,at,M}(X).$$

PROOF. The proof of part (i) is almost identical to that of Proposition 4.4 except that we use Lemma 4.15, with $K = M$, in lieu of the Davies-Gaffney estimate (2.7). A careful examination of the proof of Proposition 4.4 reveals that in fact, we did not use the full strength of the Davies-Gaffney estimates, but rather only the polynomial decay bounds provided by the case $K = M$ of Lemma 4.15.

To prove part (ii), we repeat essentially verbatim the proof of Proposition 4.13, replacing $t^2 L e^{-t^2 L}$ by $t\sqrt{L} e^{-t\sqrt{L}}$. We omit the details. $\square$

4.5. Inclusion among the spaces $H^1_{L,at,M}(X)$, $H^1_{L,\mathcal{N}_h}(X)$ and $H^1_{L,\mathcal{N}_P}(X)$. Recall the spaces $H^1_{L,at,M}(X)$, $H^1_{L,\mathcal{N}_h}(X)$ and $H^1_{L,\mathcal{N}_P}(X)$ from Definition 2.2 and from §2.6 In this section, we will prove the following result.

PROPOSITION 4.17. *Suppose $M > n_0/4$. For an operator L satisfying* (**H1**)-(**H2**) *the following inclusions hold:*

$$H^1_{L,at,M}(X) \subseteq H^1_{L,\mathcal{N}_h}(X) \quad \text{and} \quad H^1_{L,at,M}(X) \subseteq H^1_{L,\mathcal{N}_P}(X).$$

PROOF. We first prove the inclusion $H^1_{L,at,M}(X) \subseteq H^1_{L,\mathcal{N}_h}(X)$. Denote by $\mathcal{M}$ the Hardy-Littlewood maximal operator in X. For $f \in L^2(X)$, and $x \in X$, we use the Davies-Gaffney estimate (2.7) to obtain (recall (2.11) and (2.6))

$$(4.25) \quad \mathcal{N}_h f(x) \leq C \sup_{(y,t)\in\Gamma(x)} \sum_{j=0}^{\infty} \left(\frac{1}{V(y,t)} \int_{B(y,t)} |e^{-t^2 L}(f\chi_{U_j(B(y,t))})(z)|^2 d\mu(z) \right)^{1/2}$$

$$\leq C \sup_{(y,t)\in\Gamma(x)} \sum_{j=0}^{\infty} V(y,t)^{-1/2} \|e^{-t^2 L}(f\chi_{U_j(B(y,t))})\|_{L^2(B(y,t))}$$

$$\leq C \sup_{(y,t)\in\Gamma(x)} \sum_{j=0}^{\infty} V(y,t)^{-1/2}$$

$$\times \exp\left(-\frac{\text{dist}^2(U_j(B(y,t)), B(y,t))}{ct^2} \right) \|f\|_{L^2(U_j(B(y,t)))}$$

$$\leq C \sup_{x\in B(y,t)} \sum_{j=0}^{\infty} 2^{-j(n+1)/2} 2^{jn/2} \left(\frac{1}{V(2^j B(y,t))} \int_{2^j B(y,t)} |f(z)|^2 d\mu(z) \right)^{1/2}$$

$$\leq C \left[\mathcal{M}(|f|^2)(x) \right]^{1/2}.$$

Thus, we have the weak-type (2,2) bound

$$\mu\{\mathcal{N}_h f > \eta\} \leq \mu\{[\mathcal{M}(|f|^2)]^{1/2} > \eta/C\} \leq C\,\eta^{-2}\|f\|_{L^2(X)},$$

since the Hardy-Littlewood maximal theorem holds in spaces of homogeneous type.

Consequently, by Lemma 4.3 we need only to establish a uniform L^1 bound on atoms, i.e., that there exists a constant $C > 0$ such that for every $(1, 2, M)$-atom a associated to a ball B of X,

$$(4.26) \qquad \|\mathcal{N}_h a\|_{L^1(X)} \leq C.$$

We write

$$(4.27) \qquad \|\mathcal{N}_h a\|_{L^1(X)} \leq \sum_{j=0}^{10} \|\mathcal{N}_h a\|_{L^1(U_j(B))} + \sum_{j=11}^{\infty} \|\mathcal{N}_h a\|_{L^1(U_j(B))} =: I + II.$$

In concert, Kolmogorov's inequality, the Hardy-Littlewood Maximal Theorem, and the doubling property (2.2) show that for every $j = 0, 1, \ldots, 10$,

$$\begin{aligned}
\int_{U_j(B)} \mathcal{N}_h a(x)\,d\mu(x) &\leq C \int_{U_j(B)} \left[\mathcal{M}(|a|^2)(x)\right]^{1/2} d\mu(x) \\
&\leq CV(U_j(B))^{1/2}\big\||a|^2\big\|_{L^1(X)}^{1/2} \\
&\leq CV(B)^{1/2}\|a\|_{L^2(X)} \leq C,
\end{aligned}$$

which gives $I \leq C$.

To handle the second sum in (4.27), fix some number $0 < \eta < 1$ such that $\eta M - n_0/4 > 0$ and split the region over which the sup is taken in the definition of $\mathcal{N}_h a$ according to whether $t \leq c\,2^{\eta j} r_B$, or $t \geq c\,2^{\eta j} r_B$. Consider first the case $t \leq c\,2^{\eta j} r_B$. In this scenario, set

$$W_j(B) := 2^{j+3}B \backslash 2^{j-3}B, \quad R_j(B) := 2^{j+5}B \backslash 2^{j-5}B, \quad \text{and} \quad E_j(B) := (R_j(B))^c$$

for every $j \geq 11$. For $x \in U_j(B)$, $d(x, y) < t$ and $t \leq c\,2^{\eta j} r_B$ thus we have $y \in W_j(B)$. Moreover, $\operatorname{dist}(W_j(B), B) \approx C2^j r_B$ and $V(2^j B) \approx V(y, 2^j r_B)$. Then Proposition 3.1 guarantees that for $x \in U_j(B)$,

$$\begin{aligned}
\mathcal{N}_h^{(1)} a(x) &:= \sup_{\substack{(y,t)\in\Gamma(x) \\ t\leq c\,2^{\eta j} r_B}} \left(\frac{1}{V(y,t)} \int_{B(y,t)} |e^{-t^2 L} a(z)|^2 d\mu(z)\right)^{1/2} \\
&\leq C \sup_{x\in B(y,t)} \left(\frac{t}{2^j r_B}\right)^{\frac{2\eta M}{1-\eta}+\frac{n_0}{2}} \left(\frac{V(2^j B)}{V(y,t)}\right)^{1/2} \left(\frac{1}{V(2^j B)} \int_{2^j B} |a(z)|^2 d\mu(z)\right)^{1/2} \\
&\leq C2^{-2\eta M j} \left[\mathcal{M}(|a|^2)(x)\right]^{1/2}.
\end{aligned}$$

Now we treat the case $t \geq c\,2^{\eta j} r_B$. Consider the following modifications of the non-tangential maximal function: for every $f \in L^2(X)$ set

$$(4.28) \qquad \mathcal{N}_h^M f(x) := \sup_{(y,t)\in\Gamma(x)} \left(\frac{1}{V(y,t)} \int_{B(y,t)} |t^{2M} L^M e^{-t^2 L} f(z)|^2 d\mu(z)\right)^{1/2}.$$

The same argument as in (4.25) shows that $\mathcal{N}_h^M f(x) \leq C\left[\mathcal{M}(|f|^2)(x)\right]^{1/2}$, uniformly for $x \in X$. For every $x \in U_j(B)$, we use the fact that $a = L^M b$ in order to obtain

$$
\mathcal{N}_h^{(2)} a(x) \;\; := \;\; \sup_{\substack{(y,t)\in\Gamma(x) \\ t\geq c\,2^{\eta j}r_B}} \left(\frac{1}{V(y,t)} \int_{B(y,t)} |e^{-t^2 L}a(z)|^2 d\mu(z) \right)^{1/2}
$$

$$
= \;\; \sup_{\substack{(y,t)\in\Gamma(x) \\ t\geq c\,2^{\eta j}r_B}} t^{-2M} \left(\frac{1}{V(y,t)} \int_{B(y,t)} |t^{2M} L^M e^{-t^2 L}b(z)|^2 d\mu(z) \right)^{1/2}
$$

$$
\leq \;\; C 2^{-2\eta M j} r_B^{-2M} (\mathcal{N}_h^M b)(x)
$$

$$
\leq \;\; C 2^{-2\eta M j} r_B^{-2M} \left[\mathcal{M}(|b|^2)(x) \right]^{1/2}.
$$

Once again, Kolmogorov's inequality, the Hardy-Littlewood Maximal Theorem and the definition of n_0 in (2.8), give that

$$
II \;\; \leq \;\; \sum_{j=11}^{\infty} C 2^{-2\eta M j} \int_{U_j(B)} \left(\left[\mathcal{M}(|a|^2)(x) \right]^{1/2} + r_B^{-2M} \left[\mathcal{M}(|b|^2)(x) \right]^{1/2} \right) d\mu(x)
$$

$$
\leq \;\; \sum_{j=11}^{\infty} C 2^{-2\eta M j} V(2^j B)^{1/2} \left(\|a\|_{L^2(X)} + r_B^{-2M} \|b\|_{L^2(X)} \right)
$$

$$
\leq \;\; \sum_{j=11}^{\infty} C 2^{-2(\eta M - \frac{n_0}{4})j} \leq C.
$$

Altogether this shows that $\|\mathcal{N}_h a\|_{L^1(X)} \leq C$ and proves claim (4.26). As a byproduct, the inclusion $H^1_{L,at,M}(X) \subseteq H^1_{L,\mathcal{N}_h}(X)$ is also obtained.

To prove that $H^1_{L,at,M}(X) \subseteq H^1_{L,\mathcal{N}_P}(X)$, one may use the technique of Stein [St3] to bound $\mathcal{N}_P$ by $\mathcal{N}_h$ plus a controllable square function, an idea suggested to the first author in this context by P. Auscher. The details may be found in [HM], Section 7, and are omitted here. Hence, the proof of Proposition 4.17 is complete. $\qquad\square$

In closing, we note that it remains an open problem to make a direct comparison between the spaces $H^1_{L,\mathcal{N}_P}(X)$ and $H^1_{L,\mathcal{N}_h}(X)$.

CHAPTER 5

Relations between atoms and molecules

We begin with the following molecular analogue of Lemma 4.3.

LEMMA 5.1. *Fix $M \in \mathbb{N}$. Assume that T is a linear operator, or a non-negative sublinear operator, bounded on $L^2(X)$, and that for every $(1,2,M,\epsilon)$-molecule m, we have*

$$(5.1) \qquad \|Tm\|_{L^1(X)} \leq C$$

with constant C independent of m. Then T is bounded from $\mathbb{H}^1_{L,mol,M}(X)$ to $L^1(X)$, and

$$\|Tf\|_{L^1(X)} \leq C\|f\|_{\mathbb{H}^1_{L,mol,M}(X)}.$$

Consequently, as in Lemma 4.3, by density, T extends to a bounded operator from $H^1_{L,mol,M}(X)$ to $L^1(X)$.

The proof is identical to the argument in the atomic case (Lemma 4.3 above), and is omitted.

Next, we have the following result.

THEOREM 5.2. *Suppose that L satisfies* (**H1**)-(**H2**). *Let $\epsilon > 0$ and $M > n_0/4$. Then $\mathbb{H}^1_{L,mol,M}(X) \subseteq H^1_{L,S_h}(X) \cap H^2(X)$ and*

$$\|f\|_{H^1_{L,S_h}(X)} \leq C\|f\|_{\mathbb{H}^1_{L,mol,M}(X)}.$$

Before proving the theorem, we state an immediate corollary.

COROLLARY 5.3. *For all $M > n_0/4$ and $\epsilon > 0$, we have that*

$$H^1_{L,mol,M}(X) = H^1_{L,at,M}(X) = H^1_{L,S_h}(X) =: H^1_L(X).$$

PROOF OF COROLLARY 5.3. We have already shown that $H^1_{L,at,M}(X) = H^1_{L,S_h}(X)$, and that $\mathbb{H}^1_{L,at,M}(X) = H^1_{L,S_h}(X) \cap H^2(X)$, with equivalence of norms. Moreover, every $(1,2,M)$-atom is, in particular, a $(1,2,M,\epsilon)$-molecule for every $\epsilon > 0$, hence $\mathbb{H}^1_{L,at,M}(X) \subseteq \mathbb{H}^1_{L,mol,M}(X)$, with

$$\|f\|_{\mathbb{H}^1_{L,mol,M}(X)} \leq \|f\|_{\mathbb{H}^1_{L,at,M}(X)}, \qquad \forall f \in \mathbb{H}^1_{L,at,M}(X).$$

Also, by Theorem 5.2, $\mathbb{H}^1_{L,mol,M}(X) \subset H^1_{L,S_h}(X) \cap H^2(X) = \mathbb{H}^1_{L,at,M}(X)$, with

$$\|f\|_{\mathbb{H}^1_{L,at,M}(X)} \approx \|f\|_{H^1_{L,S_h}(X)} \leq C\|f\|_{\mathbb{H}^1_{L,mol,M}(X)}.$$

Consequently, $\mathbb{H}^1_{L,mol,M}(X) = H^1_{L,S_h}(X) \cap H^2(X) = \mathbb{H}^1_{L,at,M}(X)$, with equivalence of norms, so that the completions $H^1_{L,S_h}(X), H^1_{L,at,M}(X), H^1_{L,mol,M}(X)$ are all the same, independently of $M > n_0/4$, or $\epsilon > 0$. $\qquad \square$

PROOF OF THEOREM 5.2. By Lemma 5.1 and Definition 2.4, it will be enough to show that S_h maps allowable $(1, 2, M, \epsilon)$-molecules uniformly into L^1. To this end, let m be a $(1, 2, M, \epsilon)$-molecule, adapted to the ball B with radius r_B. In particular, we have that

$$(5.2) \qquad \|m\|_{L^2(X)} \leq CV(B)^{-1/2}.$$

Hence, by the doubling property and L^2 boundedness of S_h, we have that

$$\|S_h m\|_{L^1(16B)} \leq CV(B)^{1/2}\|S_h m\|_{L^2(X)} \leq C.$$

Writing now $\|S_h m\|_1 = \|S_h m\|_{L^1(16B)} + \sum_{j=5}^{\infty} \|S_h m\|_{L^1(U_j(B))}$, where we recall that $U_j(B) := 2^j B \setminus 2^{j-1} B$, we see that it is enough to prove that

$$(5.3) \qquad \|S_h m\|_{L^2(U_j(B))} \leq C 2^{-j\alpha} V(2^j B)^{-1/2},$$

for some $\alpha > 0$ and for each $j \geq 5$. To this end, we write

$$\|S_h m\|^2_{L^2(U_j(B))}$$
$$= \int_{U_j(B)} \int_0^\infty \int_{d(x,y)<t} \left|\left(t^2 L e^{-t^2 L} m\right)(y)\right|^2 \frac{d\mu(y)}{V(x,t)} \frac{dt}{t} d\mu(x)$$
$$= \int_{U_j(B)} \int_0^{2^{\theta(j-5)} r_B} \int_{d(x,y)<t} + \int_{U_j(B)} \int_{2^{\theta(j-5)} r_B}^\infty \int_{d(x,y)<t} =: I^2 + II^2,$$

where $\theta \in (0, 1)$ will be chosen momentarily. Then by Fubini's theorem, (4.2), the definition of a $(1, 2, M, \epsilon)$-molecule (Definition 2.3), and (3.4), we have for an appropriate $b \in \mathcal{D}(L^M)$,

$$II^2 \leq \int_{2^{\theta(j-5)} r_B}^\infty \int_X \left|\left(t^{2(M+1)} L^{M+1} e^{-t^2 L} b\right)(y)\right|^2 d\mu(y) \frac{dt}{t^{4M+1}}$$
$$\leq C \left(2^{\theta j} r_B\right)^{-4M} \|b\|^2_{L^2(X)} \leq C 2^{-j(4\theta M - n_0)} 2^{-j n_0} V(B)^{-1}$$
$$\leq C 2^{-j(4\theta M - n_0)} V(2^j B)^{-1},$$

where in the last step we have used (2.8). Taking square roots, and choosing θ sufficiently close to 1, we obtain (5.3) for the contribution of the term II, with $\alpha = (4\theta M - n_0)/2 > 0$.

We now treat the term I. We set

$$\widetilde{U}_j(B) := 2^{j+1} B \setminus 2^{j-2} B, \qquad \widehat{U}_j(B) := 2^{j+2} B \setminus 2^{j-3} B,$$

and observe that, by Fubini's Theorem, (4.2) and the triangle inequality

$$I^2 \le \int_0^{2^{\theta(j-5)}r_B} \int_{\widetilde{U}_j(B)} \left|\left(t^2 L e^{-t^2 L} m\right)(y)\right|^2 d\mu(y)\, \frac{dt}{t}$$

$$\lesssim \int_0^{2^{\theta(j-5)}r_B} \int_{\widetilde{U}_j(B)} \left|\left(t^2 L e^{-t^2 L}\left(1_{2^{j-3}B}m\right)\right)(y)\right|^2 d\mu(y)\, \frac{dt}{t}$$

$$+ \int_0^{2^{\theta(j-5)}r_B} \int_{\widetilde{U}_j(B)} \left|\left(t^2 L e^{-t^2 L}\left(1_{\widehat{U}_j(B)}m\right)\right)(y)\right|^2 d\mu(y)\, \frac{dt}{t}$$

$$+ \int_0^{2^{\theta(j-5)}r_B} \int_{\widetilde{U}_j(B)} \left|\left(t^2 L e^{-t^2 L}\left(1_{X\setminus 2^{j+2}B}m\right)\right)(y)\right|^2 d\mu(y)\, \frac{dt}{t}$$

$$=: (I_1)^2 + (I_2)^2 + (I_3)^2.$$

By (3.14), Definition 2.3 and the doubling property,

$$I_2 \le C\, \|m\|_{\widehat{U}_j(B)} \le C\, 2^{-j\epsilon} V(2^j B)^{-1/2},$$

which is (5.3) for the contribution of I_2. For the other two terms, we have by the generalized Davies-Gaffney estimates in Proposition 3.1,

$$(I_1)^2 + (I_3)^2 \le C\|m\|_{L^2(X)}^2 \int_0^{2^{\theta(j-5)}r_B} \exp\left(\frac{-(2^j r_B)^2}{c\,t^2}\right) \frac{dt}{t}$$

$$\le C_N \|m\|_{L^2(X)}^2 \int_0^{2^{\theta(j-5)}r_B} \left(\frac{t}{2^j r_B}\right)^N \frac{dt}{t} \le C_N V(B)^{-1} 2^{N(\theta-1)j},$$

where we have used (5.2) in the last step, and N is at our disposal. Having fixed $\theta < 1$ above, we may now choose N so large that $N(1-\theta) \ge 4M$, and then use (2.8) to obtain in turn the desired bound

$$(I_1)^2 + (I_3)^2 \le C\,V(B)^{-1} 2^{-jn_0} 2^{-j(4M-n_0)} \le C\,V(2^j B)^{-1} 2^{-j(4M-n_0)},$$

whence (5.3) follows. $\square$

Before we state the next theorem, let us make some observations. By (5.3) and its proof, we claim that an L^2 bounded linear operator mapping $(1,2,M)$ atoms into $(1,2,M,\epsilon)$ molecules for some $\epsilon > 0$, has a bounded extension from $H_L^1(X)$ into itself. Indeed, if $f = \sum \lambda_j a_j \in \mathbb{H}^1_{L,at,M}(X)$, then by the L^2 boundedness of T we have that the sum (cf. (4.5))

$$\sum \lambda_j m_j := \sum \lambda_j T(a_j) = T\left(\sum \lambda_j a_j\right)$$

converges in L^2, and is therefore a molecular $(1,2,M,\epsilon)$-representation of $T(f)$, i.e., $T(f) \in \mathbb{H}^1_{L,mol,M}(X)$ (cf. Definition 2.4). Thus, for an appropriate choice of atomic $(1,2,M)$-representation of f, we obtain

$$\|T(f)\|_{\mathbb{H}^1_{L,mol,M}(X)} \le \sum |\lambda_j| \approx \|f\|_{\mathbb{H}^1_{L,at,M}(X)}.$$

But we have observed in the proof of Corollary 5.3 that $\mathbb{H}^1_{L,mol,M}(X) = \mathbb{H}^1_{L,at,M}(X)$, whose completion is H_L^1, and the claim follows.

Secondly, if we set $\mathcal{D}^1_{L,at,M}(X)$ to be the space of all finite linear combinations of $(1, 2, M)$-atoms, i.e.

$$\mathcal{D}^1_{L,at,M}(X) = \left\{ f : f = \sum_{i=0}^{N} \lambda_i a_i, \ \lambda_i \in \mathbb{C} \ \text{and} \ a_i \ \text{are atoms} \right\},$$

then the space $\mathcal{D}^1_{L,at,M}(X)$ is a dense subspace of $H^1_{L,at,M}(X)$. In general, for every $f = \sum_{i=0}^{N} \lambda_i a_i$, with a_i $(1, 2, M)$-atoms, there exists a constant C_f such that

$$\sum_{i=1}^{N} |\lambda_i| \leq C_f \|f\|_{H^1_{L,at,M}(X)}.$$

However, C_f can not be chosen universally for all $f \in H^1_{L,at,M}(X)$ (this can be seen from Definition 2.1).

We can now state our next result.

THEOREM 5.4. *Suppose that L satisfies (H1)-(H2). Let $\epsilon = M - n_0/4 > 0$. Suppose $f = \sum_{i=0}^{N} \lambda_i a_i$, where $\{a_i\}_{i=0}^{N}$ is a family of $(1, 2, 2M)$-atoms and $\sum_{i=0}^{N} |\lambda_i| < \infty$. Then there is a representation of $f = \sum_{i=0}^{K} \mu_i m_i$, where the m_i's are $(1, 2, M, \epsilon)$-molecules and*

$$(5.4) \qquad C_1 \|f\|_{H^1_{L,at,M}(X)} \leq \sum_{i=0}^{K} |\mu_i| \leq C_2 \|f\|_{H^1_{L,at,M}(X)},$$

with $C_j = C_j(X, L, M, \epsilon, n_0)$ for $j = 1, 2$.

PROOF. Since $\{a_i\}_{i=0}^{N}$ is a family of $(1, 2, 2M)$-atoms, by definition there exist a family of functions $\{b_i\}_{i=0}^{N}$ and a family of balls $\{B_i\}_{i=0}^{N}$ such that for every $0 \leq i \leq N$, $a_i = L^{2M} b_i$ satisfies conditions (ii) and (iii) in Definition 2.1. Fix a point $x_0 \in X$. By L^2-functional calculus, for $f = \sum_{i=0}^{N} \lambda_i a_i \in H^2(X)$, we can write

$$
\begin{aligned}
f \ &= \ C_M \int_0^{\infty} (t^2 L e^{-t^2 L})^{M+2} f \frac{dt}{t} \\
&= \ C_M \int_{K_1}^{\infty} (t^2 L e^{-t^2 L})^{M+2} f \frac{dt}{t} + C_M \int_0^{1/K_2} (t^2 L e^{-t^2 L})^{M+2} f \frac{dt}{t} \\
&\quad + C_M \int_{1/K_2}^{K_1} (t^2 L e^{-t^2 L})^{M+1} \left(\left[t^2 L e^{-t^2 L} f \right] \chi_{B(x_0, K_3)^c} \right) \frac{dt}{t} \\
&\quad + C_M \int_{1/K_2}^{K_1} (t^2 L e^{-t^2 L})^{M+1} \left(\left[t^2 L e^{-t^2 L} f \right] \chi_{B(x_0, K_3)} \right) \frac{dt}{t} \\
(5.5) \qquad &= \ : f_1 + f_2 + f_3 + f_4,
\end{aligned}
$$

where the parameters K_1, K_2 and K_3 will be chosen later.

Let us start with the term f_1. Set $\mu := N^{-1} \|f\|_{H^1_{L,at,M}(X)}$. Substituting $f = \sum_{i=0}^{N} \lambda_i a_i$ into f_1, we have

$$(5.6) \qquad f_1(x) = C_M \sum_{i=0}^{N} \lambda_i \int_{K_1}^{\infty} (t^2 L e^{-t^2 L})^{M+2} a_i(x) \frac{dt}{t} = \sum_{i=0}^{N} \mu_i m_{i,K_1}(x),$$

where $\mu_i := C_M\mu$, $m_{i,K_1} := L^M f_{i,K_1}$, and

$$f_{i,K_1}(x) := \mu^{-1}\lambda_i \int_{K_1}^{\infty} t^{2(M+2)} L^2 e^{-(M+2)t^2 L} a_i(x)\frac{dt}{t}.$$

Then $\sum_{i=0}^{N} |\mu_i| = C_M \|f\|_{H^1_{L,at,M}(X)}$. We now claim that, for an appropriate choice of K_1, for every $i = 0, 1, \ldots, N$, the function m_{i,K_1} is a $(1, 2, M, \epsilon)$-molecule associated to the ball B_i. To see why this claim is true, observe that since $a_i = L^{2M} b_i$, it follows from Proposition 3.1, with $K = M + 2$, and Definition 2.1 (iii) (with $2M$ in place of M, since a_i is a $2M$ atom), that for every $k = 0, 1, \ldots, M$ and for every $i = 0, 1, \ldots, N$,

$$\left\| (r_{B_i}^2 L)^k f_{i,K_1} \right\|_{L^2(U_j(B_i))}$$

$$\leq \mu^{-1}|\lambda_i| \int_{K_1}^{\infty} t^{-2M} \left\| (t^2 L)^{2M+2} e^{-(M+2)t^2 L} (r_{B_i}^2 L)^k b_i \right\|_{L^2(U_j(B_i))} \frac{dt}{t}$$

$$\leq C\mu^{-1}|\lambda_i| \|(r_{B_i}^2 L)^k b_i\|_{L^2(B_i)} \int_{K_1}^{\infty} t^{-2M} \left(\frac{t}{2^j r_{B_i}}\right)^{\frac{n_0}{2}+\epsilon} \frac{dt}{t}$$

$$\leq C r_{B_i}^{2M} 2^{-j\epsilon} V(2^j B_i)^{-1/2} \left[\mu^{-1}|\lambda_i| \left(\frac{r_{B_i}}{K_1}\right)^{2(M-\frac{n_0}{4}-\frac{\epsilon}{2})}\right],$$

where $j = 0, 1, 2, \ldots$. This gives

$$\left\| (r_{B_i}^2 L)^k f_{i,K_1} \right\|_{L^2(U_j(B_i))} \leq r_{B_i}^{2M} 2^{-j\epsilon} V(2^j B_i)^{-1/2}$$

by choosing

$$(5.7) \qquad K_1 := C(\max_{0 \leq i \leq N} r_{B_i}) \left[\mu^{-1} \max_{0 \leq i \leq N} |\lambda_i|\right]^{\frac{1}{2(M-\frac{n_0}{4}-\frac{\epsilon}{2})}}.$$

The claim is proved.

For the term f_2, we let $\mu = N^{-1}\|f\|_{H^1_{L,at,M}(X)}$, and write

$$(5.8) \qquad f_2(x) = C_M \sum_{i=0}^{N} \lambda_i \int_0^{1/K_2} (t^2 L e^{-t^2 L})^{M+2} a_i(x)\frac{dt}{t} = \sum_{i=0}^{N} \mu_i m_{i,K_2}(x),$$

where $\mu_i := C_M\mu$, $m_{i,K_2} := L^M f_{i,K_2}$, and

$$f_{i,K_2}(x) := \mu^{-1}\lambda_i \int_0^{1/K_2} t^{2(M+2)} L^2 e^{-(M+2)t^2 L} a_i(x)\frac{dt}{t}.$$

Then $\sum_{i=0}^{N} |\mu_i| = C_M \|f\|_{H^1_{L,at,M}(X)}$. Using the condition $a_i = L^{2M} b_i$, Proposition 3.1 with $K = 2$, and Definition 2.1 (iii) (with $2M$ in place of M), we have for every $k = 0, 1, \ldots, M$, $i = 0, 1, \ldots, N$, and for every $j = 0, 1, \ldots,$

$$\left\| (r_{B_i}^2 L)^k f_{i,K_2} \right\|_{L^2(U_j(B_i))}$$
$$\leq \mu^{-1} |\lambda_i| \int_0^{1/K_2} t^{-2M} \left\| (t^2 L)^{2M+2} e^{-(M+2)t^2 L} (r_{B_i}^2 L)^k b_i \right\|_{L^2(U_j(B_i))} \frac{dt}{t}$$
$$\leq \mu^{-1} |\lambda_i| \left\| (r_{B_i}^2 L)^k b_i \right\|_{L^2(B_i)} \int_0^{1/K_2} t^{-2M} \left(\frac{t}{2^j r_{B_i}} \right)^{2M + n_0/2 + \epsilon} \frac{dt}{t}$$
$$\leq C r_{B_i}^{2M} 2^{-j\epsilon} V(2^j B_i)^{-1/2} \left[\mu^{-1} |\lambda_i| \left(r_{B_i} K_2 \right)^{-\left(\frac{n_0}{2} + \epsilon \right)} \right],$$

which gives

$$\left\| (r_{B_i}^2 L)^k f_{i,K_2} \right\|_{L^2(U_j(B_i))} \leq r_{B_i}^{2M} 2^{-j\epsilon} V(2^j B_i)^{-1/2}$$

by choosing

$$(5.9) \qquad K_2 := C \max_{0 \leq i \leq N} r_{B_i}^{-1} \cdot \left[\mu^{-1} \max_{0 \leq i \leq N} |\lambda_i| \right]^{\frac{2}{n_0 + 2\epsilon}}.$$

This shows that for each $0 \leq i \leq N$, the function m_{i,K_2} is a $(1, 2, M, \epsilon)$-molecule associated to the ball B_i.

Consider the term f_3. Let $\mu = N^{-1} \|f\|_{H^1_{L,at,M}(X)}$. One can write

$$f_3(x) = C_M \sum_{i=0}^{N} \lambda_i \int_{1/K_2}^{K_1} (t^2 L e^{-t^2 L})^{M+1} \left(\left[t^2 L e^{-t^2 L} a_i \right] \chi_{B(x_0, K_3)^c} \right)(x) \frac{dt}{t}$$

$$(5.10) \qquad = \sum_{i=0}^{N} \mu_i F_{i, K_1, K_2, K_3}(x),$$

where $\mu_i := C_M \mu$, $m_{i, K_1, K_2, K_3} := L^M f_{i, N, K_2, K_3}$, and

$$f_{i, K_1, K_2, K_3} := \mu^{-1} \lambda_i \int_{1/K_2}^{K_1} t^{2(M+1)} L e^{-(M+1)t^2 L} \left(\left[t^2 L e^{-t^2 L} a_i \right] \chi_{B(x_0, K_3)^c} \right) \frac{dt}{t}.$$

Then $\sum_{i=0}^{N} |\mu_i| = C_M \|f\|_{H^1_{L,at,M}(X)}$. We now claim that, for K_1, K_2 as above, and for an appropriate choice of K_3, the m_{i, K_1, K_2, K_3} are $(1, 2, M, \epsilon)$-molecules associated to the ball B_i. To establish the claim, for every $i = 0, 1, \ldots, N$ and every $j = 0, 1, 2, \ldots,$ we set

$$\Omega^{(1)}_{x_0, i, j, K_3} := B(x_0, K_3)^c \cap (2^{j+2} B_i \setminus 2^{j-2} B_i)$$

and

$$\Omega^{(2)}_{x_0, i, j, K_3} := B(x_0, K_3)^c \cap (2^{j+2} B_i \setminus 2^{j-2} B_i)^c.$$

Then $B(x_0, K_3)^c = \Omega^{(1)}_{x_0,i,j,K_3} \cup \Omega^{(2)}_{x_0,i,j,K_3}$ for any $j = 0, 1, 2, \ldots$. One has

$$f_{i,K_1,K_2,K_3}$$

$$= \mu^{-1}\lambda_i \int_{1/K_2}^{K_1} t^{2(M+1)} L e^{-(M+1)t^2 L} \left(\left[t^2 L e^{-t^2 L} a_i \right] \chi_{\Omega^{(1)}_{x_0,i,j,K_3}} \right) \frac{dt}{t}$$

$$+ \quad \mu^{-1}\lambda_i \int_{1/K_2}^{K_1} t^{2(M+1)} L e^{-(M+1)t^2 L} \left(\left[t^2 L e^{-t^2 L} a_i \right] \chi_{\Omega^{(2)}_{x_0,i,j,K_3}} \right) \frac{dt}{t}$$

$$=: g_{i,K_1,K_2,K_3} + h_{i,K_1,K_2,K_3}.$$

If $K_3 \geq 3 \max_{0 \leq i \leq N} d(x_0, x_{B_i})$, then for each $j = 0, 1, \ldots$, we have that $\mathrm{dist}(\Omega^{(1)}_{x_0,i,j,K_3}, B_i) \geq \max(2^{j-3} r_{B_i}, 16^{-1} K_3)$. Fix K_1 and K_2 as above. It follows from Proposition 3.1 that for any $k = 0, 1, ..., M$,

$$\left\| (r_{B_i}^2 L)^k g_{i,K_1,K_2,K_3} \right\|_{L^2(U_j(B_i))}$$

$$\leq \quad \mu^{-1}|\lambda_i| r_{B_i}^{2M} \left\| \int_0^{K_1} \left(\frac{t}{r_{B_i}} \right)^{2M-2k} (t^2 L)^{k+1} e^{-(M+1)t^2 L} \right.$$

$$\left. \left(\left[t^2 L e^{-t^2 L} a_i \right] \chi_{\Omega^{(1)}_{x_0,i,j,K_3}} \right) \frac{dt}{t} \right\|_{L^2(U_j(B_i))}$$

$$\leq \quad C\mu^{-1}|\lambda_i| r_{B_i}^{2M} \int_0^{K_1} \left(\frac{t}{r_{B_i}} \right)^{2M-k} \left\| t^2 L e^{-t^2 L} a_i \right\|_{L^2(\Omega^{(1)}_{x_0,i,j,K_3})} \frac{dt}{t}$$

$$\leq \quad C\mu^{-1}|\lambda_i| r_{B_i}^{2M} \int_0^{K_1} \left(\frac{t}{r_{B_i}} \right)^{2M-k} \left(\frac{t}{2^j r_{B_i}} \right)^{\frac{n_0}{2}+\epsilon} \left(\frac{t}{K_3} \right) \frac{dt}{t} \|a_i\|_{L^2(B_i)}$$

$$\leq \quad C r_{B_i}^{2M} 2^{-j\epsilon} V(2^j B_i)^{-1/2} \left[\mu^{-1}|\lambda_i| K_1^{2M-k+\frac{n_0}{2}+\epsilon+1} r_{B_i}^{-2M+k-\frac{n_0}{2}-\epsilon} K_3^{-1} \right].$$

The same estimate holds for the term h_{i,K_1,K_2,K_3}, since

$$\mathrm{dist}\left(U_j(B_i), \Omega^{(2)}_{x_0,i,j,K_3} \right) \geq C 2^j r_{B_i},$$

and since

$$\mathrm{dist}\left(B_i, \Omega^{(2)}_{x_0,i,j,K_3} \right) \geq c K_3$$

if $K_3 >> 2^j r_{B_i}$. Hence, we obtain that

$$\left\| (r_{B_i}^2 L)^k f_{i,K_1,K_2,K_3} \right\|_{L^2(U_j(B_i))} \leq r_{B_i}^{2M} 2^{-j\epsilon} V(2^j B_i)^{-1/2},$$

by choosing $K_3 \geq 3 \max_{0 \leq i \leq N} d(x_0, x_{B_i})$ and also

$$(5.11) \quad K_3 \geq C\mu^{-1} \max_{0 \leq k \leq M} \left(K_1^{2M-k+\frac{n_0}{2}+\epsilon+1} \max_{1 \leq i \leq N} |\lambda_i| \left[\min_{1 \leq i \leq N} r_{B_i} \right]^{-2M+k-\frac{n_0}{2}-\epsilon} \right).$$

This proves our claim.

Finally, let us consider the term f_4. Since $f \in H^1_{L,S_h}(X) \cap H^2(X)$, it follows that $F := t^2 L e^{-t^2 L} f \in T_2^1(X)$. By Proposition 4.10, F has a T_2^1-atomic decomposition:

$$(5.12) \qquad\qquad F = \sum_{i=0}^{\infty} \mu_i A_i,$$

where $\sum_{i=0}^{\infty} |\mu_i| \leq C\|F\|_{T_2^1} \leq C\|f\|_{H_{L,at,M}^1(X)}$, and A_i are T_2^1-atoms supported in $\widehat{B_i}$ satisfying $\int_{X\times(0,\infty)} |A_i(x,t)|^2 dxdt/t \leq V(B_i)^{-1}$. Substituting the decomposition (5.12) of F into f_4, we have

$$
\begin{aligned}
f_4(x) \;=\; & C_M \int_{1/K_2}^{K_1} (t^2 L e^{-t^2 L})^{M+1} \Big(\sum_{i=0}^{K_4} \mu_i A_i(\cdot,t)\chi_{B(x_0,K_3)} \Big)(x)\frac{dt}{t} \\
& + C_M \int_{1/K_2}^{K_1} (t^2 L e^{-t^2 L})^{M+1} \Big(\sum_{i=K_4+1}^{\infty} \mu_i A_i(\cdot,t)\chi_{B(x_0,K_3)} \Big)(x)\frac{dt}{t}
\end{aligned}
$$

$$
(5.13) \qquad =:\; G_{K_{1234}}(x) + H_{K_{1234}}(x),
$$

where $K_4 \in \mathbb{N}$ will be chosen in the sequel.

For the term $H_{K_{1234}}$, we let $B_0 = B(x_0,1)$ be the ball centered at x_0 and radius 1. One can write

$$
H_{K_{1234}}(x) = \|f\|_{H_{L,at,M}^1(X)} L^M h_{K_{1234}}(x)
$$

where

$$
\begin{aligned}
h_{K_{1234}}(x) := \; & \|f\|_{H_{L,at,M}^1(X)}^{-1} \\
& \times \int_{1/K_2}^{K_1} t^{2(M+1)} L e^{-(M+1)t^2 L} \Big(\sum_{i=K_4+1}^{\infty} \mu_i A_i(\cdot,t)\chi_{B(x_0,K_3)} \Big)(x)\frac{dt}{t}.
\end{aligned}
$$

Set $F_{K_4} = \sum_{i=K_4+1}^{\infty} \mu_i A_i$, and let $\eta_{K_4} = \|F_{K_4}\|_{T_2^1(X)}$. By Proposition 3.1, we have for $k = 0,1,\ldots,M$ that

$$
\begin{aligned}
\Big\| L^k h_{K_{1234}} \Big\|_{L^2(U_j(B_0))} \leq \; & C\|f\|_{H_{L,at,M}^1(X)}^{-1} \\
& \times \int_{1/K_2}^{K_1} t^{2M-2k} \Big\| (t^2 L)^{k+1} e^{-(M+1)t^2 L} (F_{K_4}\mathbf{1}_{B(x_0,K_3)})(x) \Big\|_{L^2(U_j(B_0))} \frac{dt}{t} \\
\leq \; & C_{K_3}\|f\|_{H_{L,at,M}^1(X)}^{-1} K_1^{2(M-k)} \int_{1/K_2}^{K_1} \Big(\frac{t}{2^j}\Big)^{\frac{n_0}{2}+\epsilon} \|F_{K_4}\chi_{B(x_0,K_3)}\|_{L^2(X)} \frac{dt}{t} \\
\leq \; & C_{K_3}\|f\|_{H_{L,at,M}^1(X)}^{-1} 2^{-j(\epsilon+\frac{n_0}{2})} K_1^{2(M-k)+\frac{n_0}{2}+\epsilon} K_2^{\frac{1}{2}} \\
& \times \Big(\int_{1/K_2}^{K_1} \int_{B(x_0,K_3)} |F_{K_4}(y,t)|^2 d\mu(y)dt \Big)^{1/2} \\
\leq \; & C_{K_1 K_2 K_3} 2^{-j\epsilon} V(2^j B_0)^{-1/2} \Big[V(B_0)^{1/2}\|f\|_{H_{L,at,M}^1(X)}^{-1} \eta_{K_4} \Big],
\end{aligned}
$$

where the last inequality follows by using estimate (4.11).

Note that

$$
\eta_{K_4} = \Big\| \sum_{i=K_4+1}^{\infty} \mu_i A_i \Big\|_{T_2^1(X)} \longrightarrow 0 \qquad \text{as } K_4 \to +\infty.
$$

By choosing K_4 such that η_{K_4} is sufficiently small, we have

$$\left\| L^k h_{K_{1234}} \right\|_{L^2(U_j(B_0))} \leq 2^{-j\epsilon} V(2^j B_0)^{-1/2}.$$

Therefore, $H_{K_{1234}}$ is a $(1, 2, M, \epsilon)$-molecule associated to the ball B_0.

Finally, we consider the term $G_{K_{1234}}$. For each $i = 0, 1, 2, \ldots, K_4$ we let $\widetilde{A}_i = A_i \chi_{B(x_0, K_3)}$, and observe that $\widetilde{A}_i$ is also a T_2^1-atom, supported in $\widehat{B}_i$. One can write

$$G_{K_{1234}}(x) = \sum_{i=0}^{K_4} \mu_i \left(C_M \int_{1/K_2}^{K_1} \left(t^2 L e^{-t^2 L} \right)^{M+1} \left(\widetilde{A}_i(\cdot, t) \right)(x) \frac{dt}{t} \right)$$

$$=: \sum_{i=0}^{K_4} \mu_i m_i(x)$$

Then, by a variant of Lemma 4.11, using also the Gaffney bounds for $t^2 L e^{-t^2 L}$, we obtain that, up to normalization by a multiplicative constant, for each $i = 0, 1, 2, \ldots, K_4$, the function m_i is an $(1, 2, M, \epsilon)$-molecule associated to the ball B_i. Moreover, we have

$$\sum_{i=0}^{K_4} |\mu_i| \leq \sum_{i=0}^{\infty} |\mu_i| \leq C \left\| \sum_{i=0}^{\infty} \mu_i A_i \right\|_{T_2^1(X)} \leq C \|f\|_{H^1_{L, at, M}(X)}$$

as desired. To finish the proof of Theorem 5.4, we combine the estimates we obtained for f_1, f_2, f_3 and f_4. $\square$

Theorem 5.4 yields the following immediate corollary.

COROLLARY 5.5. *Let T be a linear or positive sub-linear operator. Suppose that there is some $M > n_0/4$, and $\varepsilon \leq M - n_0/4$, for which T maps $(1, 2, M, \varepsilon)$-molecules uniformly into L^1. Then T extends by continuity to a bounded operator from $H^1_L(X)$ into $L^1(X)$.*

Analogous results in the classical setting may be found in [**MSV**], [**HZ**], [**HLZ**], [**RV**], [**CYZ**] and [**YZ**].

CHAPTER 6

BMO$_{L,M}(X)$: Duality with Hardy spaces

We start with an auxiliary lemma that gives an equivalent characterization of BMO$_{L,M}(X)$ using the resolvent of L in place of the heat semigroup associated with L. In the sequel we will frequently use the characterization below in place of the definition of BMO$_{L,M}(X)$.

LEMMA 6.1. *Let L be an operator satisfying* (**H1**)-(**H2**) *and fix $M > n_0/4$. A functional $f \in \mathcal{E}_M$ belongs to* BMO$_{L,M}(X)$ *if and only if*

$$(6.1) \qquad \sup_{B \subset X} \left(\frac{1}{V(B)} \int_B |(I - (I + r_B^2 L)^{-1})^M f(x)|^2 d\mu(x) \right)^{1/2} < \infty,$$

where the supremum is taken over all balls B in X.

PROOF. The proof is similar to that of Lemma 8.1 in [**HM**] corresponding to the case $X = \mathbb{R}^n$ and we omit the details. $\qquad \qquad \square$

THEOREM 6.2. *Let L be an operator satisfying* (**H1**)-(**H2**). *Then, for any $f \in$ BMO$_{L,M}(X)$ and $M > n_0/4$, the linear functional given by*

$$\ell(g) := \langle f, g \rangle,$$

initially defined on the dense subspace of $\mathcal{M}_0^{1,2,M,\epsilon}(L)$ consisting of finite linear combinations of $(1, 2, M, \epsilon)$-molecules, $\epsilon > 0$, and where the pairing is that between $\mathcal{M}_0^{1,2,M,\epsilon}(L)$ and its dual, has a unique bounded extension to $H_{L,at,M}^1(X)$ with

$$\|\ell\|_{(H_{L,at,M}^1(X))^*} \le C\|f\|_{\text{BMO}_{L,M}(X)}, \quad \text{for some } C \text{ independent of } f.$$

To prove Theorem 6.2, we use the following result of M. Christ ([**Ch**], Theorem 11)[1], which shows that X possesses a dyadic grid analogous to that of the Euclidean space. Specifically, we have the following.

LEMMA 6.3. *There exist a collection of open subsets $\{Q_\alpha^k \subset X : k \in \mathbb{Z}, \alpha \in I_k\}$, where I_k denotes some (possibly finite) index set depending on k, and constants $\delta \in (0,1), a_0 \in (0,1)$ and $0 < C_1 < \infty$ such that*
(i) $\mu(X \backslash \cup_\alpha Q_\alpha^k) = 0$ for all $k \in \mathbb{Z}$.
(ii) If $i \ge k$ then either $Q_\beta^i \subset Q_\alpha^k$ or $Q_\beta^i \cap Q_\alpha^k = \emptyset$.
(iii) For each (k, α) and each $i < k$, there is a unique β such that $Q_\alpha^k \subset Q_\beta^i$.
(iv) Diameter $(Q_\alpha^k) \le C_1 \delta^k$.
(v) Each Q_α^k contains some ball $B(z_\alpha^k, a_0 \delta^k)$.

---

[1]In fact, one could avoid invoking the full strength of Christ's result, but we do not pursue this point here.

PROOF OF THEOREM 6.2. Let us prove first that for every $(1, 2, M, \epsilon)$-molecule m,

$$(6.2) \qquad |\langle f, m \rangle| \leq C \|f\|_{\mathrm{BMO}_{L,M}(X)}.$$

By definition, $f \in (\mathcal{M}_0^{1,2,M,\epsilon}(L))^*$, so in particular $(I - (I + r_B^2 L)^{-1})^M f \in L^2_{\mathrm{loc}}(X)$ (see the discussion preceding (2.17)). Thus, with B denoting the ball associated with m, we may write

$$\langle f, m \rangle \;=\; \int_X (I - (I + r_B^2 L)^{-1})^M f(x)\overline{m(x)}\, dx$$
$$+ \Big\langle \Big[I - (I - (I + r_B^2 L)^{-1})^M \Big] f, m \Big\rangle$$
$$=: \; I_1 + I_2.$$

Recall (2.6). For the term I_1, we have

$$(6.3) \quad |I_1|$$

$$\leq \sum_{j=0}^{\infty} \Big(\int_{U_j(B)} |(I - (I + r_B^2 L)^{-1})^M f(x)|^2 dx \Big)^{1/2} \Big(\int_{U_j(B)} |m(x)|^2 dx \Big)^{1/2}$$

$$\leq \sum_{j=0}^{\infty} 2^{-j\epsilon} V(2^j B)^{-1/2} \Big(\int_{U_j(B)} |(I - (I + r_B^2 L)^{-1})^M f(x)|^2 dx \Big)^{1/2},$$

by Cauchy-Schwarz's inequality and the L^2-normalization of m. With notation as in Lemma 6.3, we can select an integer k_0 such that $C_1 \delta^{k_0} \leq r_B < C_1 \delta^{k_0 - 1}$ and, for each j, an integer k_j such that $\delta^{-k_j} \leq 2^j < \delta^{-k_j - 1}$. With x_B denoting the center of B, define

$$M_j := \{ \beta \in I_{k_0} : Q_\beta^{k_0} \cap B(x_B, C_1 \delta^{k_0 - k_j - 2}) \neq \emptyset \}$$

so that

$$(6.4) \qquad U_j(B) \subset B(x_B, C_1 \delta^{k_0 - k_j - 2}) \subset \bigcup_{\beta \in M_j} Q_\beta^{k_0} \subset B(x_B, 2C_1 \delta^{k_0 - k_j - 2}).$$

By Lemma 6.3, the sets $Q_\beta^{k_0}$, $\beta \in M_j$, are pairwise disjoint and for each $\beta \in M_j$ there exists $z_\beta^{k_0} \in X$ such that

$$(6.5) \qquad B(z_\beta^{k_0}, c_1 r_B) \subset Q_\beta^{k_0} \subset B(z_\beta^{k_0}, c_2 r_B)$$

for some c_1, c_2 independent of j. Hence, returning with (6.4) back to (6.3) we obtain

$$(6.6) \quad |I_1|$$

$$\leq \sum_{j=0}^{\infty} 2^{-j\epsilon} V(2^j B)^{-1/2} \Big(\sum_{\beta \in M_j} \int_{B(z_\beta^{k_0}, c_2 r_B)} |(I - (I + r_B^2 L)^{-1})^M f(x)|^2 dx \Big)^{1/2}$$

$$\leq \sum_{j=0}^{\infty} 2^{-j\epsilon} V(2^j B)^{-1/2} \|f\|_{\mathrm{BMO}_{L,M}(X)} \Big(\sum_{\beta \in M_j} V(B(z_\beta^{k_0}, c_2 r_B)) \Big)^{1/2},$$

where for the second inequality in (6.6) we used Lemma 6.1. Moreover, because of (2.4), (6.5), and (6.4) we can further write

$$\sum_{\beta \in M_j} V(B(z_\beta^{k_0}, c_2 r_B)) \;\leq\; C \sum_{\beta \in M_j} V(B(z_\beta^{k_0}, c_1 r_B)) \leq C \sum_{\beta \in M_j} V(Q_\beta^{k_0})$$

$$(6.7) \qquad\qquad \leq \; CV(B(x_B, 2C_1\delta^{k_0-k_j-2})) \leq CV(2^j B).$$

Combining (6.3), (6.6), and (6.7), we can conclude that

$$(6.8) \qquad\qquad |I_1| \leq C\|f\|_{\mathrm{BMO}_{L,M}(X)}.$$

To analyze I_2, we follow (8.15) in [**HM**] to write

$$L^M\left[I - \left(I - (1+r_B^2 L)^{-1}\right)^M\right]$$

$$= \left((r_B^{-2} + L)^M - L^M\right)\left(I - (1+r_B^2 L)^{-1}\right)^M$$

$$= \left(\sum_{k=1}^{M} \frac{M!}{(M-k)!k!} r_B^{-2k} L^{M-k}\right)\left(I - (1+r_B^2 L)^{-1}\right)^M.$$

This, together with the condition $m = L^M b$ and the fact that L is self adjoint, gives

$$|I_2| \leq C r_B^{-2M} \sum_{k=1}^{M}\left|\int_X (I - (I+r_B^2 L)^{-1})^M f(x)\overline{(r_B^2 L)^{M-k} b(x)}\,dx\right|$$

$$\leq C r_B^{-2M} \sum_{k=1}^{M}\sum_{j=0}^{\infty}\left(\int_{U_j(B)} |(I - (I+r_B^2 L)^{-1})^M f(x)|^2 dx\right)^{1/2}$$

$$\times \left(\int_{U_j(B)} |(r_B^2 L)^{M-k} b(x)|^2 dx\right)^{1/2}.$$

From this point on we proceed as in the case of I_1 using (ii) of Definition 2.3. This yields the same bound as for I_1. Given all these, (6.2) follows.

Our next goal is to show that for every number $N \in \mathbb{N}$ and for every $g = \sum_{j=0}^{N} \lambda_j a_j \in H^1_{L,at,M}(X)$, where $\{a_j\}_{j=0}^{N}$ are $(1, 2, 2M)$-atoms, we have

$$(6.9) \qquad \left|\int_X f(x)g(x)d\mu(x)\right| \leq C\|g\|_{H^1_{L,at,M}(X)}\|f\|_{\mathrm{BMO}_{L,M}(X)}.$$

Since the space of finite linear combinations of $(1, 2, 2M)$-atoms is dense in $H^1_{L,at,M}(X)$, the linear functional ℓ will then have a unique bounded extension to $H^1_{L,at,M}(X)$ defined in a standard fashion by continuity.

Let us prove claim (6.9). By Theorem 5.4, there is a representation of

$$g = \sum_{j=0}^{N} \lambda_j a_j = \sum_{i=0}^{K} \mu_i m_i,$$

where the $\{m_i\}_{i=0}^{K}$ are $(1, 2, M, \epsilon)$-molecules and

$$(6.10) \qquad\qquad \sum_{i=0}^{K} |\mu_i| \leq C\|g\|_{H^1_{L,at,M}(X)}.$$

Therefore,

$$\left| \int_X f(x)g(x)d\mu(x) \right| \leq \sum_{i=0}^{K} |\mu_i| \left| \int_X f(x)m_i(x)d\mu(x) \right|$$

$$\leq C \sum_{i=0}^{K} |\mu_i| \|f\|_{\mathrm{BMO}_{L,M}(X)}$$

$$\leq C\|g\|_{H^1_{L,at,M}(X)} \|f\|_{\mathrm{BMO}_{L,M}(X)}.$$

This proves claim (6.9), which in turn finishes the proof of Theorem 6.2. $\square$

Our next result is essentially the converse of Theorem 6.2.

THEOREM 6.4. *Let $M > n_0/4$ and $\epsilon > 0$. Suppose that L satisfies* (**H1**)-(**H2**) *and that ℓ is a bounded linear functional on $H^1_{L,at,M}(X)$. Then in fact, $\ell \in \mathrm{BMO}_{L,M}(X)$ and for all $g \in H^1_{L,at,M}(X)$ which can be represented as finite linear combinations of $(1,2,M,\epsilon)$-molecules, there holds*

$$\ell(g) = \langle f, g \rangle$$

where the pairing is that between $\mathcal{M}_0^{1,2,M,\epsilon}(L)$ and its dual. Moreover,

$$\|\ell\|_{\mathrm{BMO}_{L,M}(X)} \leq C\|\ell\|_{(H^1_{L,at,M}(X))^*}.$$

PROOF. By Theorem 5.2, we have that for any $(1,2,M,\epsilon)$-molecule m,

$$\|m\|_{H^1_{L,at,M}(X)} \leq C.$$

Thus,

$$|\ell(m)| \leq C\|\ell\|_{(H^1_{L,at,M}(X))^*}$$

for every $(1,2,M,\epsilon)$-molecule m. In particular, $\ell \in \mathcal{E}_M$ for every $M > n_0/4$. Thus, $(I - (I + r_B^2 L)^{-1})^M \ell$ is well defined and belongs to $L^2_{\mathrm{loc}}(X)$ (see the discussion preceding (2.17)). Fix a ball B, and let $\varphi \in L^2(B)$, with $\|\varphi\|_{L^2(B)} \leq 1$. As we observed before,

$$\widetilde{m} := V(B)^{-1/2}(I - (I + r_B^2 L)^{-1})^M \varphi$$

is (up to a multiplicative constant) a $(1,2,M,\epsilon)$-molecule. Thus,

$$V(B)^{-1/2} |\langle (I - (I + r_B^2 L)^{-1})^M \ell, \varphi \rangle|$$

$$= V(B)^{-1/2} |\langle \ell, (I - (I + r_B^2 L)^{-1})^M \varphi \rangle|$$

$$= |\langle \ell, \widetilde{m} \rangle| \leq C\|\ell\|_{(H^1_{L,at,M}(X))^*}.$$

Taking the supremum over all such φ supported in B, we obtain

$$\frac{1}{V(B)} \int_B |(I - (I + r_B^2 L)^{-1})^M \ell(x)|^2 dx \leq C \left(\|\ell\|_{(H^1_{L,at,M}(X))^*} \right)^2.$$

Finally, taking the supremum over all balls B in X, the conclusion of the theorem follows. $\square$

In concert, Theorem 6.2, Theorem 6.4, and Corollary 5.3, justify Theorem 2.7 stated in Section 2.

CHAPTER 7

Hardy spaces and Gaussian estimates

In this section, we give further results about Hardy spaces $H^1_{L,at,M}(X)$ assuming some "Gaussian" upper bounds for the heat kernel of the operator L. More precisely, we assume that:

(H1) L is a non-negative self-adjoint operator on $L^2(X)$;

(H3) The kernel of e^{-tL}, denoted by $p_t(x,y)$, is a measurable function on $X \times X$ and there exist two positive constants C and c such that, for almost every $x, y \in X$,

$$(7.1) \qquad |p_t(x,y)| \leq \frac{C}{V(x,\sqrt{t})} \exp\left(-\frac{d^2(x,y)}{ct}\right), \qquad \forall\, t > 0.$$

We note that obviously **(H3)** implies **(H2)**. It is also useful to note that Gaussian upper bounds for $p_t(x,y)$ are further inherited by the time derivatives of $p_t(x,y)$. That is, for each $k \in \mathbb{N}$, there exist two positive constants c_k and C_k such that

$$(7.2) \qquad \left|\frac{\partial^k}{\partial t^k} p_t(x,y)\right| \leq \frac{C_k}{t^k V(x,\sqrt{t})} \exp\left(-\frac{d^2(x,y)}{c_k t}\right), \qquad \forall\, t > 0,$$

for almost every $x, y \in X$. For the proof of (7.2), see **[CD2]**, **[Da3]**, **[Gr]** and **[Ou]**, Theorem 6.17.

7.1. Hardy spaces $H^1_{L,at,M}(X)$, $H^1_{L,S_h}(X)$ and $H^1_{L,S_P}(X)$ and Gaussian estimates. In this subsection we establish certain improved versions of Propositions 4.4 and 4.13, under the stronger assumption that Gaussian upper bounds hold.

THEOREM 7.1. *If an operator L satisfies conditions* **(H1)** *and* **(H3)**, *then for every number $M \geq 1$, the spaces $H^1_{L,at,M}(X)$ and $H^1_{L,S_h}(X)$ coincide. In particular,*

$$\|f\|_{H^1_{L,at,M}(X)} \approx \|f\|_{H^1_{L,S_h}(X)}.$$

Remark: In the context of Hodge Laplacians, it has already been observed in **[AMR]** that it suffices to take $M \geq 1$ in the presence of pointwise Gaussian bounds.

PROOF. As in Chapter 4, it is enough to work with the dense spaces $\mathbb{H}^1_{L,at,M}(X)$ and $H^1_{L,S_h}(X) \cap H^2(X)$. The inclusion $H^1_{L,S_h}(X) \cap H^2(X) \subseteq \mathbb{H}^1_{L,at,M}(X)$ was proved in Proposition 4.13, for every $M \geq 1$, and does not require Gaussian estimates. As for the converse inclusion, by Lemma 4.3 it suffices to verify that for any $(1,2,1)$-atom a associated to a ball $B = B(x_B, r_B)$, there holds

$$(7.3) \qquad \|S_h a\|_{L^1(X)} \leq C.$$

Since S_h is bounded on $L^2(X)$, by Hölder's inequality we have that

$$\|S_h a\|_{L^1(4B)} \leq C.$$

45

Going further, we use the fact that $a = Lb$ for some $b \in \mathcal{D}(L)$ satisfying (ii) and (iii) in Definition 2.1. For $x \notin 4B$, this allows us to write

$$(S_h a(x))^2 = \int_0^\infty \int_{d(x,y)<t} \left| t^2 L e^{-t^2 L} a(y) \right|^2 \frac{d\mu(y)}{V(x,t)} \frac{dt}{t}$$

$$= \int_0^{r_B} \int_{d(x,y)<t} \left| t^2 L e^{-t^2 L} a(y) \right|^2 \frac{d\mu(y)}{V(x,t)} \frac{dt}{t}$$

$$+ \left(\int_{r_B}^{d(x,x_B)/4} + \int_{d(x,x_B)/4}^\infty \right) \int_{d(x,y)<t} \left| (t^2 L)^2 e^{-t^2 L} b(y) \right|^2 \frac{d\mu(y)}{V(x,t)} \frac{dt}{t^5}$$

$$=: E_1(x) + E_2(x) + E_3(x).$$

It follows from (2.4) and (2.5) that for $x \notin 4B$, $z \in B$, and $0 < t < d(x,x_B)/4$, we have

$$(7.4) \qquad V(z,t)^{-1} \le C V(x_B, d(x,x_B))^{-1} \left(\frac{d(x,x_B)}{t} \right)^{n_0}.$$

On the other hand, if $x \notin 4B$ and $d(x,y) < t$, we also have $d(x,y) \le t < d(x,x_B)/4$ and $d(y,z) \ge d(x,x_B)/2$ for every $z \in B$. These, together with (4.2), the fact that $-\frac{d}{dt}[e^{-tL}] = L e^{-tL}$, and (7.2) for $k = 1$, show that

$$(7.5) \quad E_1(x)$$

$$\le C \int_0^{r_B} \int_{d(x,y)<t} \left| \int_B \frac{1}{V(z,t)} \exp\left(-\frac{d^2(y,z)}{ct^2} \right) |a(z)| d\mu(z) \right|^2 \frac{d\mu(y)}{V(x,t)} \frac{dt}{t}$$

$$\le C d(x,x_B)^{2n_0} V(x_B, d(x,x_B))^{-2} \|a\|_{L^1(B)}^2 \int_0^{r_B} t^{-2n_0} \left(\frac{t}{d(x,x_B)} \right)^{2(n_0+1)} \frac{dt}{t}$$

$$\le \frac{C}{V(x_B, d(x,x_B))^2} \frac{r_B^2}{d(x,x_B)^2}.$$

The second inequality in (7.5) makes use of (7.4) and the fact that $d(y,z) \ge d(x,x_B)/2$. A similar argument shows that

$$E_2(x)$$

$$\le C d(x,x_B)^{2n_0} V(x_B, d(x,x_B))^{-2} \|b\|_{L^1(B)}^2 \int_{r_B}^\infty t^{-2n_0} \left(\frac{t}{d(x,x_B)} \right)^{2n_0+2} \frac{dt}{t^5}$$

$$\le \frac{C}{V(x_B, d(x,x_B))^2} \frac{r_B^2}{d(x,x_B)^2}.$$

Consider the term $E_3(x)$. For $z \in B$ and $t \ge d(x,x_B)/4$, we have that $V(z,t)^{-1} \le C V(x_B, d(x,x_B))^{-1}$ for $x \notin 4B$. This, together with estimate (7.2) for $k = 1$, gives

$$E_3(x) \le C V(x_B, d(x,x_B))^{-2} \|b\|_{L^1(B)}^2 \int_{d(x,x_B)/4}^\infty \frac{dt}{t^5}$$

$$\le \frac{C}{V(x_B, d(x,x_B))^2} \frac{r_B^2}{d(x,x_B)^2}.$$

The estimates obtained for $E_1(x), E_2(x)$ and $E_3(x)$ combine to show that $\|S_h a\|_{L^1((4B)^c)} \leq C$. This justifies (7.3), and the proof of Theorem 7.1 is completed. $\qquad\square$

Turning to the equivalence between $H^1_{L,at,M}(X)$ and $H^1_{L,S_P}(X)$, we start with the following auxiliary result.

LEMMA 7.2. *For every $K = 0, 1, \ldots$, there exists a constant C_K such that the kernel $p_{t,K}$ of the operator $(t\sqrt{L})^{2K} e^{-t\sqrt{L}}$ satisfies*

$$|p_{t,K}(x,y)| \leq C_K \frac{1}{V(x,t)} \left(1 + \frac{d(x,y)}{t}\right)^{-(2K+1)} \qquad \forall\, t > 0,$$

for almost every $x, y \in X$.

PROOF. The subordination formula (4.22) allows us to estimate

$$|p_{t,K}(x,y)| \leq C \int_0^\infty \frac{e^{-u}}{\sqrt{u}} \cdot \frac{1}{V(x, \frac{t}{2\sqrt{u}})} \exp\left(-\frac{u d^2(x,y)}{c_K t^2}\right) u^K du.$$

Note that there exists $C > 0$ such that

$$(7.6) \qquad e^{-\frac{u}{2}} V\left(x, \frac{t}{2\sqrt{u}}\right)^{-1} \leq C V(x,t)^{-1} \quad \forall\, u > 0,\ x \in X,\ t > 0.$$

Indeed, if $0 < u < 1/4$, this is true for trivial reasons (with $C = 1$), whereas if $u \geq 1/4$, from the doubling property (2.4) we have

$$e^{-\frac{u}{2}} \frac{1}{V(x, \frac{t}{2\sqrt{u}})} \leq C e^{-\frac{u}{2}} \left(\sqrt{u}\right)^{n_0} \frac{1}{V(x,t)} \leq \frac{C}{V(x,t)}.$$

Therefore, using (7.6),

$$\begin{aligned}
|p_{t,K}(x,y)| &\leq \frac{C}{V(x,t)} \int_0^\infty e^{-\frac{u}{2}} \exp\left(-\frac{u d^2(x,y)}{c_K t^2}\right) u^{K-1/2} du \\
&\leq \frac{C}{V(x,t)} \left(1 + \frac{d(x,y)}{t}\right)^{-(2K+1)}
\end{aligned}$$

for every $t > 0$ and almost every $x, y \in X$. $\qquad\square$

With Lemma 7.2 in hand, by reasoning as in the proof of Theorem 7.1, one then obtains the following result.

THEOREM 7.3. *If an operator L satisfies conditions* **(H1)** *and* **(H3)***, then for every number $M \geq 1$, the spaces $H^1_{L,at,M}(X)$ and $H^1_{L,S_P}(X)$ coincide. In particular,*

$$\|f\|_{H^1_{L,at,M}(X)} \approx \|f\|_{H^1_{L,S_P}(X)}.$$

7.2. Hardy spaces via maximal functions. We continue the discussion from subsection 4.4 regarding the study of Hardy spaces in terms of maximal and non-tangential maximal functions, under the additional assumption that Gaussian upper bounds hold.

Given an operator L satisfying **(H1)** and **(H3)** (stated in the first part of Section 7) and a function $f \in L^2(X)$, consider the following maximal and non-tangential maximal operators associated to the heat semigroup generated by the operator L,

$$(7.7) \qquad f_h^+(x) := \sup_{t>0} |e^{-t^2 L} f(x)|,$$

and

$$(7.8) \qquad N_h f(x) := \sup_{d(x,y)<t} |e^{-t^2 L} f(y)|.$$

In addition, let us also consider the Poisson semigroup generated by the operator L and the operators

$$(7.9) \qquad f_P^+(x) := \sup_{t>0} |e^{-t\sqrt{L}} f(x)|$$

and

$$(7.10) \qquad N_P f(x) := \sup_{d(x,y)<t} |e^{-t\sqrt{L}} f(y)|$$

with $x \in X, f \in L^2(X)$.

Define the spaces $H^1_{L,\max,h}(X)$, $H^1_{L,N_h}(X)$, $H^1_{L,\max,P}(X)$ and $H^1_{L,N_P}(X)$ as the completion of $H^2(X)$ in the norms given by the $L^1(X)$ norm of the corresponding maximal function. For example,

$$\|f\|_{H^1_{L,\max,h}(X)} = \|f_h^+\|_{L^1(X)},$$

etc. Then the following theorem holds.

THEOREM 7.4. *If an operator L satisfies conditions* (**H1**) *and* (**H3**), *then for every number $M \geq 1$, the following continuous inclusions hold:*
 (i) $H^1_{L,at,M}(X) \subseteq H^1_{L,N_h}(X) \subseteq H^1_{L,\max,h}(X) \subseteq H^1_{L,\max,P}(X);$
 (ii) $H^1_{L,at,M}(X) \subseteq H^1_{L,N_P}(X).$

Remark: It is trivial that the "averaged" non-tangential maximal functions $\mathcal{N}_P$ and $\mathcal{N}_h$ are dominated by their pointwise analogues N_P and N_h, respectively, so that $H^1_{L,N_P}(X) \subseteq H^1_{L,\mathcal{N}_P}(X)$ and $H^1_{L,N_h}(X) \subseteq H^1_{L,\mathcal{N}_h}(X)$, but it is not clear how to reverse these inclusions, nor to compare $H^1_{L,\mathcal{N}_P}(X)$ and $H^1_{L,\mathcal{N}_h}(X)$ to their pointwise "vertical" analogues, $H^1_{L,\max,P}(X)$ and $H^1_{L,\max,h}(X)$, in the absence of some sort of "Moser-type" local boundedness (cf. Lemma 8.4 below).

PROOF. We first prove the inclusion $H^1_{L,at,M}(X) \subseteq H^1_{L,N_h}(X)$. By Lemma 4.3, it suffices to show that there exists $C > 0$ such that for every atom a associated to a ball $B = B(x_B, r_B)$ in X, we have

$$(7.11) \qquad \|N_h a\|_{L^1(X)} \leq C.$$

The condition (7.1) implies that $N_h a(x) \leq C\mathcal{M}a(x)$ for almost everywhere $x \in X$, where $\mathcal{M}$ denotes the Hardy-Littlewood maximal operator on X. By Hölder's inequality, we then have

$$\|N_h a\|_{L^1(4B)} \leq V(4B)^{1/2}\|\mathcal{M}a\|_{L^2(X)} \leq CV(B)^{1/2}\|a\|_{L^2(B)} \leq C.$$

For $x \notin 4B$, the same type of argument as in Theorem 7.1 shows that if $d(x,y) \leq t < d(x,x_B)/4$ and $z \in B$, then $d(y,z) \geq d(x,x_B)/2$, and

$$V(z,t)^{-1} \leq CV(x_B, d(x,x_B))^{-1}\left(\frac{d(x,x_B)}{t}\right)^n.$$

We now estimate $N_h a(x)$ with $x \notin 4B$ by examining several cases. To facilitate the subsequent presentations, introduce for $f \in L^2(X)$,

$$(7.12) \qquad N_h^{(1)} f(x) \;=\; \sup_{\substack{d(x,y)<t \\ 0<t\leq r_B}} |e^{-t^2 L} f(y)|,$$

$$(7.13) \qquad N_h^{(2)} f(x) \;=\; \sup_{\substack{d(x,y)<t \\ r_B<t<d(x,x_B)/4}} |e^{-t^2 L} f(y)|,$$

$$(7.14) \qquad N_h^{(3)} f(x) \;=\; \sup_{\substack{d(x,y)<t \\ t\geq d(x,x_B)/4}} |e^{-t^2 L} f(y)|.$$

Case 1. $0 < t \leq r_B$. In this scenario we have

$$N_h^{(1)} a(x) \leq C \sup_{\substack{d(x,y)<t \\ 0<t\leq r_B}} \int_B V(z,t)^{-1} \exp\left(-\frac{d(y,z)^2}{ct^2} \right) |a(z)| d\mu(z)$$

$$\leq \sup_{0<t\leq r_B} \frac{C}{V(x_B, d(x,x_B))} \left(\frac{d(x,x_B)}{t} \right)^{n_0} \exp\left(-\frac{d(x,x_B)^2}{ct^2} \right) \|a\|_{L^1(B)}$$

$$\leq \frac{C}{V(x_B, d(x,x_B))} \frac{r_B}{d(x,x_B)}.$$

Case 2. $r_B < t < d(x,x_B)/4$. Since a is a $(1,2,1)$-atom, we can write $a = Lb$ for some $b \in \mathcal{D}(L)$ satisfying (ii) and (iii) in Definition 2.1. Then, one has

$$N_h^{(2)} a(x) = \sup_{\substack{d(x,y)<t \\ r_B<t<d(x,x_B)/4}} t^{-2} |t^2 L e^{-t^2 L} b(y)|$$

$$\leq C \sup_{\substack{d(x,y)<t \\ r_B<t<d(x,x_B)/4}} t^{-2} \int_B V(z,t)^{-1} \exp\left(-\frac{d(y,z)^2}{ct^2} \right) |b(z)| d\mu(z)$$

$$\leq \frac{C}{V(x_B, d(x,x_B))} \|b\|_{L^1(B)}$$

$$\times \sup_{r_B<t<d(x,x_B)/4} t^{-2} \left(\frac{d(x,x_B)}{t} \right)^{n_0} \exp\left(-\frac{d(x,x_B)^2}{ct^2} \right)$$

$$\leq \frac{C}{V(x_B, d(x,x_B))} \frac{r_B}{d(x,x_B)}.$$

Case 3. $t \geq d(x,x_B)/4$. In this case, $V(z,t)^{-1} \leq CV(x_B, d(x,x_B))^{-1}$ for every $z \in B$, and then

$$N_h^{(3)} a(x) = \sup_{\substack{d(x,y)<t \\ t \geq d(x,x_B)/4}} |Le^{-t^2 L} b(y)|$$

$$\leq C \sup_{t \geq d(x,x_B)/4} t^{-2} V(x_B, d(x,x_B))^{-1} \|b\|_{L^1(B)}$$

$$\leq \frac{C}{V(x_B, d(x,x_B))} \frac{r_B}{d(x,x_B)}.$$

Combining the estimates obtained in *Case 1*, *Case 2* and *Case 3*, we may conclude that

$$(7.15) \qquad N_h a(x) \leq \frac{C}{V(x_B, d(x,x_B))} \frac{r_B}{d(x,x_B)}.$$

Integrating both sides of (7.15) over $X \setminus 4B$ yields (7.11). Thus, the proof of the continuous inclusion $H^1_{L,at,M}(X) \subseteq H^1_{L,N_h}(X)$ is justified.

The proof of $H^1_{L,N_h}(X) \subseteq H^1_{L,\max,h}(X)$ follows from the definitions of the maximal operators N_h and of the f_h^+. Moreover, we have

$$(7.16) \qquad \|f_h^+\|_{L^1(X)} \leq \|N_h f\|_{L^1(X)}.$$

Next, we prove the inclusion $H^1_{L,\max,h}(X) \subseteq H^1_{\max,P}(X)$. To do so, fix $f \in H^1_{L,\max,h}(X)$. The subordination formula (4.22) can be used to estimate

$$|e^{-t\sqrt{L}} f(x)| \leq C \int_0^\infty \frac{e^{-u}}{\sqrt{u}} \left| e^{-\frac{t^2}{4u} L} f(x) \right| du$$

$$\leq C f_h^+(x) \int_0^\infty \frac{e^{-u}}{\sqrt{u}} du$$

$$\leq C f_h^+(x).$$

This proves that $f_P^+(x) \leq C f_h^+(x)$ for any $x \in X$. Thus, $H^1_{L,\max,h}(X) \subseteq H^1_{L,\max,P}(X)$.

The proof of the inclusion $H^1_{L,at,1}(X) \subseteq H^1_{L,N_P}(X)$ is similar to that of $H^1_{L,at,1}(X) \subseteq H^1_{L,N_h}(X)$ and we omit the details. This completes the proof of Theorem 7.4. $\qquad \square$

Parenthetically we remark that it seems likely that under additional assumptions, such as Nash type local Hölder continuity of the heat kernel, one may obtain equality of the various spaces in Theorem 7.4. We do not attempt to address this point here, but see [**AR**].

7.3. The spaces $\mathrm{BMO}_L(X)$ under Gaussian bounds. Call a function $f \in L^2_{\mathrm{loc}}(X)$ of type (x_0, β), where $x_0 \in X$ and $\beta > 0$, if it satisfies

$$(7.17) \qquad \|f\|_{\mathcal{M}_{(x_0,\beta)}} := \left(\int_X \frac{|f(x)|^2}{(1+d(x_0,x))^\beta V(x_0, 1+d(x_0,x))} d\mu(x) \right)^{1/2} < \infty.$$

We denote by $\mathcal{M}_{(x_0,\beta)}$ the collection of all functions of type (x_0, β). For a fixed $x_0 \in X$, it is easy to see that $\mathcal{M}_{(x_0,\beta)}$ is a Banach space under the norm (7.17). Also, for any $x_1 \in X$, we have $\mathcal{M}_{(x_1,\beta)} = \mathcal{M}_{(x_0,\beta)}$ with equivalent norms. Set

$$\mathcal{M} := \bigcup_{x_0 \in X} \bigcup_{\beta:\, 0<\beta<\infty} \mathcal{M}_{(x_0,\beta)}$$

We will say that $f \in \mathcal{M}$ is in $\mathrm{BMO}_L(X)$, the space of functions of bounded mean oscillation associated with $\{e^{-tL}\}_{t>0}$, if

$$(7.18) \qquad \|f\|_{\mathrm{BMO}_L(X)} := \sup_B \frac{1}{V(B)} \int_B |f(x) - e^{-r_B^2 L} f(x)| \, d\mu(x) < \infty,$$

where the supremum is taken over all balls B in X. Note that this formally corresponds to Definition 2.6 with $M = 1$ and with an L^1 norm in place of the L^2 norm (see [**DY1**], [**DY2**]). The current presence of stronger (point-wise) bounds allows us to take $M = 1$.

Next, define

$$K_L := \left\{ f \in \mathcal{M} : e^{-tL} f(x) = f(x) \text{ for almost all } x \in X \text{ and all } t > 0 \right\}.$$

We have that $K_L = \{f \in \mathrm{BMO}_L(X) : \|f\|_{\mathrm{BMO}_L(X)} = 0\}$ and $\mathrm{BMO}_L(X)/K_L$ is a Banach space with the norm

$$\|f\|_{\mathrm{BMO}_L(X)/K_L} = \|f\|_{\mathrm{BMO}_L(X)}.$$

We remark that the convention made after Definition 2.6 applies here with $M = 1$ and that the two versions of the BMO are compatible whenever they can simultaneously be considered.

The following result holds.

THEOREM 7.5. *Assume that the operator L satisfies conditions* (**H1**) *and* (**H3**). *Then, we have*

$$(H^1_{L,at,1}(X))^* = \mathrm{BMO}_L(X).$$

PROOF. This result can be proved in a similar, but slightly simpler, fashion to the duality results in Section 6. We omit the proof. $\square$

CHAPTER 8

Hardy spaces associated to Schrödinger operators

In this section we treat Hardy spaces adapted to a Schrödinger operator in $\mathbb{R}^n$, assuming merely local integrability and non-negativity of the potential. Our work extends some of the previous theory developed in [**DZ1, DZ2, DGMTZ**] under stronger hypotheses on the potential. In particular, [**DZ1**] and [**DGMTZ**] deal with a Schrödinger operator whose potential belongs to the reverse Hölder class RH_q, with $q \geq n/2$, while [**DZ2**] generalizes the results of [**DZ1**].

Let $n \geq 1$ and V be a locally integrable non-negative function on $\mathbb{R}^n$, not identically zero. We define the form Q by

$$(8.1) \qquad Q(u,v) := \int_{\mathbb{R}^n} \nabla u \nabla v \, dx + \int_{\mathbb{R}^n} V uv \, dx$$

with domain

$$(8.2) \qquad \mathcal{D}(Q) := \left\{ u \in W^{1,2}(\mathbb{R}^n) : \int_{\mathbb{R}^n} V|u|^2 dx < \infty \right\}.$$

It is well known that this symmetric form is closed. Note also that it was shown by Simon [**Sim**] that this form coincides with the minimal closure of the form given by the same expression but defined on $C_0^\infty(\mathbb{R}^n)$ (the space of C^∞ functions with compact supports). In other words, $C_0^\infty(\mathbb{R}^n)$ is a core of the form Q.

Let us denote by L the self-adjoint operator associated with Q. The domain of L is given by

$$(8.3) \qquad \mathcal{D}(L) :=$$

$$\left\{ u \in \mathcal{D}(Q) : \exists v \in L^2 \text{ such that } Q(u,\varphi) = \int_{\mathbb{R}^n} v\bar{\varphi} \, dx, \ \forall \varphi \in \mathcal{D}(Q) \right\}.$$

Formally, we write $L = -\Delta + V$ as a Schrödinger operator with potential V. Since V is a locally integrable non-negative function on $\mathbb{R}^n$, the Feynman-Kac formula implies that the semigroup kernels $p_t(x,y)$, associated to e^{-tL}, satisfy the estimates

$$(8.4) \qquad 0 \leq p_t(x,y) \leq (4\pi t)^{-\frac{n}{2}} \exp\left(-\frac{|x-y|^2}{4t} \right) \quad \text{for all } t > 0, \ x,y \in \mathbb{R}^n.$$

See page 195 of [**Ou**]. In particular, as we have noted above (see the discussion immediately before and after (2.15)), this fact implies that $H_L^2(X) = L^2(X)$.

8.1. Equivalences among $H_{L,at,M}^1(\mathbb{R}^n)$, $H_{L,S_h}^1(\mathbb{R}^n)$ and $H_{L,S_P}^1(\mathbb{R}^n)$. The following theorem is a special case of Theorems 7.1 and 7.3.

THEOREM 8.1. *Assume that $L = -\Delta + V$, where $V \in L_{\mathrm{loc}}^1(\mathbb{R}^n)$ is a non-negative function on $\mathbb{R}^n$. Then for all $M \geq 1$, the spaces $H_{L,at,M}^1(\mathbb{R}^n)$, $H_{L,S_h}^1(\mathbb{R}^n)$*

53

and $H^1_{L,S_P}(\mathbb{R}^n)$ *coincide. In particular,*

$$\|f\|_{H^1_{L,at,M}(\mathbb{R}^n)} \approx \|f\|_{H^1_{L,S_h}(\mathbb{R}^n)} \approx \|f\|_{H^1_{L,S_P}(\mathbb{R}^n)}.$$

8.2. Maximal characterization of $H^1_{L,at,M}(\mathbb{R}^n)$. In this section, we continue with the assumption that $V \in L^1_{\mathrm{loc}}(\mathbb{R}^n)$ is a non-negative function on $\mathbb{R}^n$. In the sequel, we may sometimes use capital letters to denote points in $\mathbb{R}^{n+1}_+$, e.g., $Y = (y,t)$, and set

$$\nabla_Y u(Y) = (\nabla_y u, \partial_t u) \quad \text{and} \quad |\nabla_Y u|^2 = |\nabla_y u|^2 + |\partial_t u|^2.$$

For every function $f \in L^2(\mathbb{R}^n)$, consider the quadratic operators associated to the heat semigroup and the Poisson semigroup generated by the operator L,

$$\widetilde{S}_h f(x) = \left(\iint_{\Gamma(x)} |t\nabla_Y e^{-t^2 L} f(y)|^2 \frac{dydt}{t^{n+1}} \right)^{1/2},$$

and

$$\widetilde{S}_P f(x) = \left(\iint_{\Gamma(x)} |t\nabla_Y e^{-t\sqrt{L}} f(y)|^2 \frac{dydt}{t^{n+1}} \right)^{1/2},$$

where $x \in \mathbb{R}^n, f \in L^2(\mathbb{R}^n)$.

Define the spaces $H^1_{L,\widetilde{S}_h}(\mathbb{R}^n)$ and $H^1_{L,\widetilde{S}_P}(\mathbb{R}^n)$ as the completion of $H^2(\mathbb{R}^n)$ in the norms given by the L^1 norm of the square function, e.g.,

$$\|f\|_{H^1_{L,\widetilde{S}_h}(\mathbb{R}^n)} = \|\widetilde{S}_h f\|_{L^1(\mathbb{R}^n)}.$$

Then the following result holds.

THEOREM 8.2. *Assume that $L = -\Delta + V$, where $V \in L^1_{\mathrm{loc}}(\mathbb{R}^n)$ is a non-negative function on $\mathbb{R}^n$. Then all of the Hardy spaces $H^1_{L,at,M}(\mathbb{R}^n)$, $M \geq 1$, $H^1_{L,\widetilde{S}_h}(\mathbb{R}^n)$, $H^1_{L,\widetilde{S}_P}(\mathbb{R}^n)$, $H^1_{L,\max,h}(\mathbb{R}^n)$, $H^1_{L,N_h}(\mathbb{R}^n)$, $H^1_{L,\max,P}(\mathbb{R}^n)$ and $H^1_{L,N_P}(\mathbb{R}^n)$ coincide. In other words, $\forall M \geq 1$,*

$$\|f\|_{H^1_{L,at,M}(\mathbb{R}^n)} \approx \|f\|_{H^1_{L,\widetilde{S}_h}(\mathbb{R}^n)} \approx \|f\|_{H^1_{L,\widetilde{S}_P}(\mathbb{R}^n)} \approx \|f\|_{H^1_{L,\max,h}(\mathbb{R}^n)}$$

$$\approx \|f\|_{H^1_{L,N_h}(\mathbb{R}^n)} \approx \|f\|_{H^1_{L,\max,P}(\mathbb{R}^n)} \approx \|f\|_{H^1_{L,N_P}(\mathbb{R}^n)}.$$

Remark: We note that similar equivalences hold for the spaces $H^1_{L,\mathcal{N}_P}(X)$ and $H^1_{L,\mathcal{N}_h}(X)$, by virtue of the local boundedness estimates in Lemma 8.4 below (in the case of the Poisson extension), or the analogous parabolic estimates in [**CarMSp**] (in the case of the heat extension); we leave the routine details to the interested reader.

The proof of Theorem 8.2 will be given below; we first prove some preliminary estimates.

8.2.1. *Estimates for weak solutions.* In order to prove Theorem 8.2, we need some estimates for the Poisson semigroup $\{e^{-t\sqrt{L}}\}_{t>0}$, and for weak solutions of the equation

$$\widetilde{L}u := -\Delta_{x,t}u + Vu = -u_{tt} + Lu = 0$$

in domains $\Omega \subset \mathbb{R}^{n+1}$. To define the latter notion, we suppose that Ω is an open subset of $\mathbb{R}^{n+1}$. Define

$$W^{1,2}_V(\Omega) = \left\{ u \in W^{1,2}(\Omega) : \int_\Omega V|u|^2 \, dydt < \infty \right\},$$

and let $W^{1,2}_{V,0}(\Omega)$ denote the subspace of $W^{1,2}_V(\Omega)$ with trace 0 on $\partial\Omega$. The function $u \in W^{1,2}_V(\Omega)$ is called a *weak solution* of $\widetilde{L}u = 0$ in Ω if it satisfies

$$(8.5) \qquad \int_\Omega \nabla u \cdot \nabla \varphi \, dY + \int_\Omega V u \varphi \, dY = 0 \quad \text{for every } \varphi \in W^{1,2}_{V,0}(\Omega).$$

We note that here, and in the sequel when working in an $(n+1)$-dimensional context, ∇ denotes the full gradient ∇_Y in $\mathbb{R}^{n+1}$. Moreover, the local estimates that we are about to prove (Lemmas 8.3 and 8.4) are valid for potentials V which may depend on all the variables $Y = (y,t)$. However, our results for semigroups will of course require t-independence of V.

We begin by stating a Caccioppoli inequality which appears previously in [**Sh**]. We include the proof here for the sake of self-containment.

LEMMA 8.3. *Let u be a weak solution of $\widetilde{L}u = 0$ in the ball $B(Y_0, 2r) \subset \mathbb{R}^{n+1}$. Then there exists an absolute constant $C > 0$ such that*

$$\int_{B(Y_0,r)} |\nabla u(Y)|^2 dY \leq \frac{C}{r^2} \int_{B(Y_0,2r)} |u(Y)|^2 dY.$$

PROOF. Let $\eta \in C^1_0(B(Y_0, 2r))$ with $\eta = 1$ on $B(Y_0, r)$ and $|\nabla \eta| \leq r^{-1}$. Set $\varphi = \eta^2 u$. Then we have

$$\int |\nabla u|^2 \eta^2 \, dY + \int 2\eta u \nabla u \cdot \nabla \eta \, dY = -\int V \eta^2 u^2 \, dY \leq 0.$$

This gives

$$\int |\nabla u|^2 \eta^2 dY \leq 2 \int \eta |u| |\nabla \eta| |\nabla u| dY \leq \epsilon \int |\nabla u|^2 \eta^2 dY + \frac{1}{\epsilon} \int |u|^2 |\nabla \eta|^2 dY,$$

where in the last step we have used a variant of Cauchy's inequality. Choosing $\epsilon = 1/2$, hiding the small term on the left hand side, and using the bound for $|\nabla \eta|$, we obtain Caccioppoli's inequality in the usual way. $\qquad\square$

Next, we recall a Moser type local boundedness estimate, which has appeared previously in [**AB**]. We include the proof here for the sake of self-containment.

LEMMA 8.4. *Let $u, B(Y_0, 2r)$ be as in Lemma 8.3. Then for any $p > 0$, there exits a constant $C = C(n,p) > 0$ such that*

$$\sup_{B(Y_0,r)} |u(Y)| \leq C \Big(\frac{1}{r^{n+1}} \int_{B(Y_0,2r)} |u(Y)|^p dY \Big)^{1/p}.$$

PROOF. It is enough to show that u^2 is a subharmonic. To this end, observe that for any $\varphi \in C^1_0(B(Y_0, 2r))$ with $\varphi \geq 0$, we have

$$\int \nabla u^2 \cdot \nabla \varphi \, dY$$

$$= 2 \int \nabla u \cdot \nabla (u\varphi) \, dy \, dY - 2 \int \varphi |\nabla u|^2 \, dY$$

$$= -2 \int V \varphi u^2 \, dY - 2 \int \varphi |\nabla u|^2 \, dY$$

$$\leq 0.$$

The desired result follows readily. $\qquad\square$

Next, we observe that the heat semigroup associated to L satisfies a Davies-Gaffney estimate.

LEMMA 8.5. *There exist two constants c, $C > 0$ such that for any two closed sets E and F of $\mathbb{R}^n$, we have:*

$$\|t\nabla e^{-t^2 L} f\|_{L^2(F)} \leq C e^{-\frac{\operatorname{dist}(E,F)^2}{ct^2}} \|f\|_{L^2(E)}$$

for every $f \in L^2(\mathbb{R}^n)$ supported in E.

PROOF. The proof is similar to that of the case when L is a divergence form operator, and is omitted. See, for example, Lemma 2.1 of [**AHLMT**]. $\square$

8.2.2. *Proof of Theorem 8.2.* **Step I**: Proof of the inclusion $H^1_{L,\max,P}(\mathbb{R}^n) \subseteq H^1_{L,N_P}(\mathbb{R}^n)$.
We apply Lemma 8.4 with $0 < p < 1$ and $u(x,t) = e^{-t\sqrt{L}} f(x)$ to obtain that for every $x \in \mathbb{R}^n$ and every $(y,t) \in \Gamma(x)$,

$$
\begin{aligned}
|e^{-t\sqrt{L}} f(y)|^p &\leq \frac{C}{t^{n+1}} \int_{t/2}^{2t} \int_{|x-z|<2t} |e^{-s\sqrt{L}} f(z)|^p dz\, ds \\
&\leq \frac{C}{t^n} \int_{|x-z|<2t} |f_P^+|^p(z)\, dz \leq CM\big(|f_P^+|^p\big)(x),
\end{aligned}
$$

where $\mathcal{M}$ is the Hardy-Littlewood maximal function in $\mathbb{R}^n$. We then have

$$N_h f(x) \leq C\big[\mathcal{M}\big(|f_P^+|^p\big)(x)\big]^{1/p} \quad \text{for } x \in \mathbb{R}^n.$$

Therefore, since $p < 1$,

$$\|N_h f\|_{L^1(\mathbb{R}^n)} \leq C\big\|\big[\mathcal{M}\big(|f_P^+|^p\big)\big]^{1/p}\big\|_{L^1(\mathbb{R}^n)} \leq C\|f_P^+\|_{L^1(\mathbb{R}^n)}.$$

This proves that $H^1_{L,\max,P}(\mathbb{R}^n) \subseteq H^1_{L,N_P}(\mathbb{R}^n)$.
Step II. Proof of $H^1_{L,N_P}(\mathbb{R}^n) \subseteq H^1_{L,S_P}(\mathbb{R}^n)$. The proof follows the analogous argument for the case $V = 0$ of Fefferman and Stein [**FS**], with some modifications owing to the lack of pointwise bounds for the gradient, as in, e.g., [**AR, HM, AMR**]. First, we define an area functional using all partial derivatives of $e^{-t\sqrt{L}} f(x)$ by setting

$$(8.6) \qquad \widetilde{S}_P^\beta f(x) := \Big(\iint_{\Gamma_\beta(x)} |t\nabla_Y e^{-t\sqrt{L}} f(y)|^2 \frac{dy\, dt}{t^{n+1}}\Big)^{1/2}, \qquad \text{for } x \in \mathbb{R}^n,$$

where $Y = (y,t)$. For simplicity we will write $\widetilde{S}_P f$ in place of $\widetilde{S}_P^1 f$. It is clear that $S_P f \leq \widetilde{S}_P^1 f$ pointwise in $\mathbb{R}^n$. We also define the family of truncated cones

$$\Gamma^{\epsilon,R,\alpha}(x) := \Big\{(y,t) \in \mathbb{R}^n \times (\epsilon, R) : |x - y| < \alpha t\Big\}, \qquad x \in \mathbb{R}^n,$$

and, associated to these new truncated cones, the area functions

$$\widetilde{S}_P^{\epsilon,R,\beta} f(x) := \Big(\iint_{\Gamma^{\epsilon,R,\beta}(x)} |t\nabla_Y e^{-t\sqrt{L}} f(y)|^2 \frac{dy dt}{t^{n+1}}\Big)^{1/2}, \qquad x \in \mathbb{R}^n.$$

In what follows we will work with $\widetilde{S}_P^{\epsilon,R,\beta}$ rather than $\widetilde{S}_P^\beta$ and then pass to the limit as $\epsilon \to 0$, $R \to \infty$. In the sequel, unless explicitly stated, the constants appearing in estimates will not depend on ϵ and R.

We also define the following non-tangential maximal function

$$N_P^\beta f(x) := \sup_{(y,t)\in\Gamma^\beta(x)} |e^{-t\sqrt{L}} f(y)|.$$

For every closed set $E \subseteq \mathbb{R}^n$, define

$$(8.7) \qquad E^* :=$$

$$\left\{ x \in \mathbb{R}^n : \frac{|E \cap B(x)|}{|B(x)|} \geq \frac{1}{2} \text{ for every } B(x), \text{ ball in } \mathbb{R}^n \text{ centered at } x \right\},$$

the set of points having global density bigger than or equal to $1/2$ with respect to E. For $\beta > 0$ to be selected later, we introduce

$$\mathcal{R}^{\epsilon,R,\beta}(E^*) := \bigcup_{x \in E^*} \Gamma^{\epsilon,R,\beta}(x)$$

(which is a Lipschitz domain given that it has the uniform cone property) and

$$u(y,t) := e^{-t\sqrt{L}} f(y), \qquad t \in (0,\infty), \quad y \in \mathbb{R}^n.$$

Making use of Lemma 2.1 in [**CMS**], it is not hard to see that

$$\int_{E^*} \left(\widetilde{S}_P^{2\epsilon,R,1/2} f(x) \right)^2 dx \;\leq\; \int_{E^*} \left(\widetilde{S}_P^{\alpha\epsilon,\alpha R,1/\alpha} f(x) \right)^2 dx$$

$$(8.8) \hspace{4cm} \leq\; \iint_{\mathcal{R}^{\alpha\epsilon,\alpha R,1/\alpha}(E^*)} t|\nabla_Y u(y,t)|^2 \, dy dt,$$

for all $\alpha \in (1,2)$. Going further, integration by parts shows that

$$\iint_{\mathcal{R}^{\alpha\epsilon,\alpha R,1/\alpha}(E^*)} t|\nabla_Y u(y,t)|^2 \, dy \, dt$$

$$= \iint_{\mathcal{R}^{\alpha\epsilon,\alpha R,1/\alpha}(E^*)} t\nabla_Y u(y,t) \cdot \overline{\nabla_Y u(y,t)} \, dy \, dt$$

$$= \int_{\partial\mathcal{R}^{\alpha\epsilon,\alpha R,1/\alpha}(E^*)} t\nabla_Y u(y,t) \cdot N_E(y,t)\overline{u(y,t)} \, d\sigma_E(y,t)$$

$$- \iint_{\mathcal{R}^{\alpha\epsilon,\alpha R,1/\alpha}(E^*)} t\left(\partial_t^2 u(y,t) + \triangle_y u(y,t)\right)\overline{u(y,t)} \, dy \, dt$$

$$- \iint_{\mathcal{R}^{\alpha\epsilon,\alpha R,1/\alpha}(E^*)} \partial_t u(y,t)\overline{u(y,t)} \, dy \, dt,$$

where $N_E(y,t)$ is the outward unit normal vector to $\mathcal{R}^{\alpha\epsilon,\alpha R,1/\alpha}(E^*)$ and $d\sigma_E$ is the surface measure over $\partial\mathcal{R}^{\alpha\epsilon,\alpha R,1/\alpha}(E^*)$. Observe that $\partial_t^2 u + \triangle_y u = Vu$ in the weak sense on $\mathbb{R}^{n+1}_+$. Since $0 \leq V \in L^1_{\mathrm{loc}}(\mathbb{R}^n)$, we then have

$$(8.9) \qquad \iint_{\mathcal{R}^{\alpha\epsilon,\alpha R,1/\alpha}(E^*)} t|\nabla_Y u(y,t)|^2 dy\, dt$$

$$\leq \mathrm{Re} \int_{\partial\mathcal{R}^{\alpha\epsilon,\alpha R,1/\alpha}(E^*)} t\nabla_Y u(y,t) \cdot N_E(y,t)\overline{u(y,t)}\, d\sigma_E(y,t)$$

$$- \mathrm{Re} \iint_{\mathcal{R}^{\alpha\epsilon,\alpha R,1/\alpha}(E^*)} \partial_t u(y,t)\overline{u(y,t)}\, dy\, dt,$$

where $\mathrm{Re}\, z$ denotes the real part of a complex number z. Using integration by parts again, we have

$$2\mathrm{Re} \iint_{\mathcal{R}^{\alpha\epsilon,\alpha R,1/\alpha}(E^*)} \partial_t u(y,t)\overline{u(y,t)}dy dt$$

$$= \int_{\partial\mathcal{R}^{\alpha\epsilon,\alpha R,1/\alpha}(E^*)} |u(y,t)|^2 N_E(y,t) \cdot (0,\cdots,0,1)\, d\sigma_E(y,t).$$

Using this back in (8.9) and, after taking absolute values, integrating both sides of the resulting inequality with respect to α we obtain

$$\int_1^2 \iint_{\mathcal{R}^{\alpha\epsilon,\alpha R,1/\alpha}(E^*)} t|\nabla_Y u(y,t)|^2 \, dy\, dt\, d\alpha$$

$$\leq \int_1^2 \int_{\partial\mathcal{R}^{\alpha\epsilon,\alpha R,1/\alpha}(E^*)} t|u(y,t)||\nabla_Y u(y,t)|d\sigma_E(y,t)\, d\alpha$$

$$+ \int_1^2 \int_{\partial\mathcal{R}^{\alpha\epsilon,\alpha R,1/\alpha}(E^*)} |u(y,t)|^2 d\sigma_E(y,t)\, d\alpha$$

$$\leq \iint_{\mathcal{B}^{\epsilon,R}(E^*)} |u(y,t)||\nabla_Y u(y,t)|dy\, dt + \iint_{\mathcal{B}^{\epsilon,R}(E^*)} |u(y,t)|^2 \frac{dy\, dt}{t}$$

$$\leq \left(\iint_{\mathcal{B}^{\epsilon,R}(E^*)} t|\nabla_Y u(y,t)|^2 dy\, dt\right)^{1/2} \left(\iint_{\mathcal{B}^{\epsilon,R}(E^*)} |u(y,t)|^2 \frac{dy\, dt}{t}\right)^{1/2}$$

$$(8.10) \qquad + \iint_{\mathcal{B}^{\epsilon,R}(E^*)} |u(y,t)|^2 \frac{dy\, dt}{t},$$

where

$$\mathcal{B}^{\epsilon,R}(E^*) := \left\{(x,t) \in \mathbb{R}^{n+1}_+ : (x,t) \in \partial\mathcal{R}^{\alpha\epsilon,\alpha R,1/\alpha}(E^*) \text{ for some } 1 < \alpha < 2\right\}.$$

Consider the following three regions:

$$(8.11) \qquad \mathcal{B}^{\epsilon}(E^*) := \left\{(x,t) \in \mathbb{R}^n \times (\epsilon, 2\epsilon) : \mathrm{dist}(x, E^*) < t\right\},$$

$$(8.12) \qquad \mathcal{B}^{R}(E^*) := \left\{(x,t) \in \mathbb{R}^n \times (R, 2R) : \mathrm{dist}(x, E^*) < t\right\},$$

$$(8.13) \qquad \mathcal{B}_{\epsilon,R}(E^*) := \left\{(x,t) \in \mathbb{R}^n \times (\epsilon, 2R) : \mathrm{dist}(x, E^*) < t < 2\,\mathrm{dist}(x, E^*)\right\},$$

and observe that

$$\mathcal{B}^{\epsilon,R}(E^*) \subseteq \mathcal{B}^{\epsilon}(E^*) \cup \mathcal{B}^{R}(E^*) \cup \mathcal{B}_{\epsilon,R}(E^*).$$

Below we will analyze separately the parts of integrals in (8.10) corresponding to the regions (8.11)–(8.13).

Let us start with

$$I_1 := \iint_{\mathcal{B}^\epsilon(E^*)} |u(y,t)|^2 \frac{dy\,dt}{t}.$$

For every $(y,t) \in \mathcal{B}^\epsilon(E^*)$, there exists $y^* \in E^*$ such that $y^* \in B(y,t)$. By definition of E^* this implies that $|E \cap B(y^*,t)| > |B(y^*,t)|/2$ and therefore $|E \cap B(y,2t)| \geq Ct^n$. Then

$$
\begin{aligned}
I_1 \;\leq\;& C \iint_{\mathcal{B}^\epsilon(E^*)} \int_{E \cap B(y,2t)} |u(y,t)|^2 dz\, \frac{dy\,dt}{t^{n+1}} \\
\leq\;& C \int_\epsilon^{2\epsilon} \int_E \left(t^{-n} \int_{B(z,2t)} |u(y,t)|^2 dy \right) dz\, \frac{dt}{t} \\
\leq\;& C \int_\epsilon^{2\epsilon} \int_E |N_P^\beta f(z)|^2 \frac{dz\,dt}{t} \leq C \int_E |N_P^\beta f(z)|^2 dz
\end{aligned}
$$

provided $\beta \geq 2$. Next, using similar ideas, we may estimate

$$
\begin{aligned}
I_2 \;=\;& \iint_{\mathcal{B}^\epsilon(E^*)} t |\nabla_Y u(y,t)|^2 dy\,dt \\
\leq\;& C \int_\epsilon^{2\epsilon} \int_E \left(\frac{1}{t^{n-2}} \int_{B(z,2t)} |\nabla_Y u(y,t)|^2 dy \right) \frac{dz\,dt}{t} \\
\end{aligned}
$$

$$(8.14) \qquad\qquad \leq\; C\epsilon^{1-n} \int_E \left(\int_{\mathcal{G}_z} |\nabla_Y u(y,t)|^2 dy\,dt \right) dz,$$

where $\mathcal{G}_z$ is the set of points (y,t) with $|y-z| < 2t$, $\epsilon < t < 2\epsilon$ and $z \in E$. Pick a covering of $\mathcal{G}_z$ with bounded overlap by a finite number K of balls $B_j = B((x_j, t_j), \epsilon/4)$, where $(x_j, t_j) \in \mathcal{G}_z$. That is, $\mathcal{G}_z \subseteq \cup_{j=0}^K B_j$, and every point $(y,t) \in \mathcal{G}_z$ belongs to at most a finite number of balls B_j. By geometric considerations, it follows that $\cup_{j=0}^K B((x_j, t_j), \epsilon/2) \subseteq \{(y,t) : |y-z| < 16t, \epsilon/2 < t < 3\epsilon\}$. We then apply Lemma 8.3 to obtain the bound

$$
\begin{aligned}
I_2 \;\leq\;& C \sum_{j=0}^K \epsilon^{1-n} \int_E \left(\int_{B((x_j, t_j), \epsilon/4)} |\nabla_Y u(y,t)|^2 dy\,dt \right) dz \\
\leq\;& C \sum_{j=0}^K \epsilon^{1-n} \int_E \left(\epsilon^{-2} \int_{B((x_j, t_j), \epsilon/2)} |u(y,t)|^2 dy\,dt \right) dz \\
\leq\;& C\epsilon^{-(n+1)} \int_E \left(\int_{\epsilon/2}^{3\epsilon} \int_{B(z,16t)} |u(y,t)|^2 dy\,dt \right) dz \leq C \int_E |N_P^\beta f(z)|^2 dz,
\end{aligned}
$$

where it was assumed that $\beta \geq 16$. Observe that the same argument applies to further estimate

$$\iint_{\mathcal{B}^R(E^*)} |u(y,t)|^2 \frac{dy\,dt}{t} \leq C \int_E |N_P^\beta f(z)|^2 dz,$$

$$\iint_{\mathcal{B}^R(E^*)} t |\nabla_Y u(y,t)|^2 dy\,dt \leq C \int_E |N_P^\beta f(z)|^2 dz,$$

granted that $\beta \geq 16$.

To control the integral over $\mathcal{B}_{\epsilon,R}(E^*)$, we first decompose $(E^c)^*$ into a family of Whitney balls, $\{B(x_k, r_k)\}_{k=0}^\infty$, such that $\bigcup_{k=0}^\infty B(x_k, r_k) = (E^c)^*$, $c_1 \operatorname{dist}(x_k, E^*) \leq r_k \leq c_2 \operatorname{dist}(x_k, E^*)$, and every point $x \in (E^c)^*$ belongs to at most c_3 balls. Here,

$0 < c_1 < c_2 < 1$ and $c_3 \in \mathbb{N}$ are some fixed constants, independent of $(E^c)^*$ (see
[**CW2**] and [**St1**]). Then

$$
\begin{aligned}
I_3 \ &:= \ \iint_{\mathcal{B}_{\epsilon,R}(E^*)} |u(y,t)|^2 \frac{dy\,dt}{t} \\
&\leq \ \sum_{k=0}^{\infty} \int_{r_k(1/c_2-1)}^{2r_k(1/c_1+1)} \int_{B(x_k,r_k)} |u(y,t)|^2 \frac{dy\,dt}{t}
\end{aligned}
$$

$$
(8.15) \qquad \leq \ C \sum_{k=0}^{\infty} r_k^n \int_{r_k(1/c_2-1)}^{2r_k(1/c_1+1)} \left[\frac{1}{t^n} \int_{B(x_k,\frac{c_2}{1-c_2}t)} |u(y,t)|^2 dy \right] \frac{dt}{t}.
$$

From the fact that $E^* \subseteq E$ it follows that $\operatorname{dist}(x_k,E) \leq \operatorname{dist}(x_k,E^*) \leq \frac{c_2}{(1-c_2)c_1}t$
which, in turn, can be used to majorize the term under right-most inner integral
in (8.15) by $N_P^\beta f(z)$ for some $z \in E$ provided $\beta \geq \frac{c_2}{(1-c_2)c_1}$. Hence, assuming that
this is the case,

$$
(8.16) \qquad I_3 \ \leq \ C \sum_{k=0}^{\infty} r_k^n \left(\sup_{z\in E} N_P^\beta f(z) \right)^2 \leq C|(E^c)^*| \left(\sup_{z\in E} N_P^\beta f(z) \right)^2.
$$

A reasoning similar to the one used to prove (8.15) and (8.16) also yields that
there exists a finite constant $C_0 = C_0(c_1,c_2) > 0$ such that

$$
I_4 \ := \ \iint_{\mathcal{B}_{\epsilon,R}(E^*)} t|\nabla_Y u(y,t)|^2 dy\,dt \leq C|(E^c)^*| \left(\sup_{z\in E} N_P^\beta f(z) \right)^2,
$$

provided $\beta > C_0$. Let us now choose

$$
(8.17) \qquad \beta := \max\left\{ 16, \frac{c_2}{(1-c_2)c_1}, C_0 \right\}
$$

in (8.18). Combining all the estimates above allows us to write

$$
\int_{E^*} \left(\widetilde{S}_P^{2\epsilon,R,1/2} f(x) \right)^2 dx \ \leq \ C|(E^c)^*| \left(\sup_{z\in E} N_P^\beta f(z) \right)^2 + C \int_E |N_P^\beta f(z)|^2 dz
$$

and, therefore, passing to the limit as $\epsilon \to 0$ and $R \to \infty$,

$$
\int_{E^*} \left(\widetilde{S}_P^{1/2} f(x) \right)^2 dx \ \leq \ C|(E^c)^*| \left(\sup_{z\in E} N_P^\beta f(z) \right)^2 + C \int_E |N_P^\beta f(z)|^2 dz.
$$

At this stage choose

$$
(8.18) \qquad E := \{ x \in \mathbb{R}^n : N_P^\beta f(x) \leq \sigma \},
$$

for some arbitrary, fixed $\sigma > 0$. Also, generically, denote by λ_F the distribution
function of F. Then, since $N_P^\beta f \leq \sigma$ on E and $|(E^c)^*| \leq C|E^c| \leq C\lambda_{N_P^\beta f}(\sigma)$, we
have

$$
\int_{E^*} \left(\widetilde{S}_P^{1/2} f(x) \right)^2 dx \ \leq \ C\sigma^2 \lambda_{N_P^\beta f}(\sigma) + C \int_0^\sigma t\lambda_{N_P^\beta f}(t)\,dt.
$$

Next,

$$
\begin{aligned}
\lambda_{\widetilde{S}_P^{1/2} f}(\sigma) \ &\leq \ \left| \{ x \in E^* : \widetilde{S}_P^{1/2} f(x) > \sigma \} \right| + |(E^*)^c| \\
&\leq \ C\frac{1}{\sigma^2} \int_{E^*} \left(\widetilde{S}_P^{1/2} f(x) \right)^2 dx + C\lambda_{N_P^\beta f}(\sigma) \\
&\leq \ C\frac{1}{\sigma^2} \int_0^\sigma t\lambda_{N_P^\beta f}(t)\,dt + C\lambda_{N_P^\beta f}(\sigma).
\end{aligned}
$$

Therefore, for β as in (8.17), by applying Lemma 4.6 we obtain

$$\|S_P f\|_{L^1(\mathbb{R}^n)} \leq \|\widetilde{S}_P f\|_{L^1(\mathbb{R}^n)} \;\leq\; C\|\widetilde{S}_P^{1/2} f\|_{L^1(\mathbb{R}^n)} = \int_0^\infty \lambda_{\widetilde{S}_P^{1/2} f}(\sigma)\, d\sigma$$

$$\leq\; C\int_0^\infty \lambda_{N_P^\beta f}(\sigma)\, d\sigma \leq C\|N_P^\beta f\|_{L^1(\mathbb{R}^n)}$$

$$\leq\; C\beta^n \|N_P f\|_{L^1(\mathbb{R}^n)},$$

where for the last inequality we have used the fact that $\|N_P^\beta f\|_{L^1(\mathbb{R}^n)} \leq C\beta^n \|N_P f\|_{L^1(\mathbb{R}^n)}$ (see Theorem 2.3 of [**CT**]). Hence, the inclusion

$$H^1_{L,N_P}(\mathbb{R}^n) \subseteq H^1_{L,S_P}(\mathbb{R}^n)$$

is proved, and then the proof of Theorem 8.2 is complete. $\qquad\square$

8.3. $H^1_{L,at,M} \to H^1$ bounds for Riesz transforms of Schrödinger operators. Let L be as in Section 8.1, that is $L = -\Delta + V$, with $0 \leq V \in L^1_{\mathrm{loc}}(\mathbb{R}^n)$. Consider the Riesz transform $T := \nabla L^{-1/2}$ associated to the operator L. An alternative definition is

$$(8.19) \qquad Tf = \frac{1}{2\sqrt{\pi}} \int_0^\infty \nabla e^{-tL} f \frac{dt}{\sqrt{t}}.$$

Then the operator T is bounded on $L^2(\mathbb{R}^n)$. Indeed, for every $f \in L^2(\mathbb{R}^n)$, we have

$$(8.20)\quad \|Tf\|^2_{L^2(\mathbb{R}^n} = \int_{\mathbb{R}^n} |\nabla L^{-1/2} f(x)|^2 dx$$

$$\leq \int_{\mathbb{R}^n} |\nabla L^{-1/2} f(x)|^2 dx + \int_{\mathbb{R}^n} V(x)|L^{-1/2} f(x)|^2 dx$$

$$= Q(L^{-1/2} f, L^{-1/2} f) = \|f\|^2_{L^2(\mathbb{R}^n)},$$

where Q is as in (8.1). The latter equality follows from the fact that Q is symmetric hence $\mathcal{D}(Q) = \mathcal{D}(L^{1/2})$ and $Q(u,v) = (L^{1/2}u, L^{1/2}v)$ (see, for examples, Chapter VI of [**K**] or p. 254, Theorem 8.1 of [**Ou**]).

Moreover, by the molecular decomposition of functions in $H^1_{L,S_h}(\mathbb{R}^n)$, it was proved in [**DOY**] that T is bounded from the Hardy space $H^1_{L,S_h}(\mathbb{R}^n)$ into $L^1(\mathbb{R}^n)$ and, by interpolation, T is bounded on $L^p(\mathbb{R}^n)$ for all $1 < p \leq 2$. See also [**Si2**]. We note that for $p > 2$, the counter-example studied in [**Sh**] with the potential $V(x) = |x|^{\epsilon-2}$ shows that the operator $T = \nabla(-\Delta + V)^{-1/2}$ is not necessarily bounded on L^p. However, L^p-boundedness of Riesz transforms for large values of p can be obtained if one imposes certain additional regularity conditions on the potential V (see [**Sh**], [**AB**]). Below we observe that the target space $L^1(\mathbb{R}^n)$ can be replaced by the smaller space $H^1(\mathbb{R}^n)$.

THEOREM 8.6. *Assume that $L = -\Delta + V$, where $V \in L^1_{\mathrm{loc}}(\mathbb{R}^n)$ is a non-negative function on $\mathbb{R}^n$. Then the Riesz transform $\nabla L^{-1/2}$ is bounded from $H^1_{L,at,M}(\mathbb{R}^n)$ with $M \geq 1$ (and, hence, from any of the equivalent spaces from Theorem 8.2) into $H^1(\mathbb{R}^n)$, i.e., there exists a constant $C > 0$ such that*

$$(8.21) \qquad \|\nabla L^{-1/2} f\|_{H^1(\mathbb{R}^n)} \leq C\|f\|_{H^1_{L,at,M}(\mathbb{R}^n)}.$$

PROOF. Recall that m is a molecule for $H^1(\mathbb{R}^n)$ centered at x_0 if

$$(8.22) \qquad \left\{ \int_{\mathbb{R}^n} |m(x)|^2 dx \right\}\left\{ \int_{\mathbb{R}^n} |m(x)|^2 |x - x_0|^{2n} dx \right\} \leq 1$$

and

$$(8.23) \qquad \int_{\mathbb{R}^n} m(x)\, dx = 0.$$

As is well-known (cf. Theorem C in [**CW2**]) if m is a molecule for $H^1(\mathbb{R}^n)$ centered at x_0, then $m \in H^1(\mathbb{R}^n)$ and $\|m\|_{H^1(\mathbb{R}^n)}$ depends only on n.

Fix $M > n$ (by Theorem 8.2, the choice of $M \geq 1$ does not affect the nature of the atomic space). In order to prove (8.21), it suffices to prove that for every $(1, 2, M)$-atom a associated to a ball B of $\mathbb{R}^n$, $m = \nabla L^{-1/2}a$ is a molecule. Clearly (8.23) holds. Since $\nabla L^{-1/2}$ is bounded on $L^2(\mathbb{R}^n)$, we have

$$(8.24) \qquad \|m\|_{L^2(\mathbb{R}^n)} = \|\nabla L^{-1/2}a\|_{L^2(\mathbb{R}^n)} \leq C\|a\|_{L^2(B)} \leq C|B|^{-1/2}.$$

We now estimate $\|\,|x - x_B|^n m\|_{L^2(\mathbb{R}^n)}$. Since a is a $(1, 2, M)$-atom, there exists a function $b \in \mathcal{D}(L^M)$ such that $a = L^M b$, which satisfies conditions (ii) and (iii) in Definition 2.1. We can write

$$\left\|\,|x - x_B|^n m\right\|_{L^2(\mathbb{R}^n)} \leq \sum_{j=0}^{\infty} \left\|\,|x - x_B|^n m\right\|_{L^2(U_j(B))} \leq \sum_{j=0}^{\infty}(2^j r_B)^n \|m\|_{L^2(U_j(B))}.$$

The formula

$$L^{-1/2}a = \frac{1}{\sqrt{\pi}}\int_0^{\infty} e^{-t^2 L}a\, dt,$$

allows us to write for $j = 2, 3, \ldots,$

$$\begin{aligned}
\|m\|_{L^2(U_j(B))} &= \frac{1}{\sqrt{\pi}}\left\|\int_0^{\infty} \nabla e^{-t^2 L}a\, dt\right\|_{L^2(U_j(B))} \\
&\leq C\int_0^{r_B} \|t\nabla e^{-t^2 L}a\|_{L^2(U_j(B))}\frac{dt}{t} \\
&\quad + C\int_{r_B}^{\infty} t^{-2M}\|t\nabla(t^2 L)^M e^{-t^2 L}b\|_{L^2(U_j(B))}\frac{dt}{t} \\
&= : I + II.
\end{aligned}$$

To estimate I, we use Lemma 8.5 to obtain

$$\begin{aligned}
(8.25) \qquad I &\leq C\int_0^{r_B}\exp\left(-\frac{2^{2j}r_B}{ct^2}\right)\frac{dt}{t}\|a\|_{L^2(B)} \\
&\leq C\int_0^{r_B}\left(\frac{t}{2^j r_B^2}\right)^{n+1}\frac{dt}{t}|B|^{-1/2} \leq C2^{-j(n+1)}|B|^{-1/2}.
\end{aligned}$$

Consider the term II. Again we use Lemma 8.5 and Lemma 2.3 of [**HMa**] (which essentially says that the composition of two operators satisfying Davies-Gaffney estimates also satisfies Davies-Gaffney estimates) to write

$$(8.26) \qquad \begin{aligned} II \ &\leq \ \int_{r_B}^{\infty} t^{-2M-1} \| t\nabla e^{-\frac{t^2}{2}L}(t^2 L)^M e^{-\frac{t^2}{2}L} b \|_{L^2(U_j(B))} dt \\ &\leq \ C\|b\|_{L^2(B)} \int_{r_B}^{\infty} t^{-2M-1} \exp\Big(-\frac{2^{2j}r_B^2}{ct^2}\Big) dt \\ &\leq \ Cr_B^{2M}|B|^{-1/2} \int_{r_B}^{\infty} t^{-2M-1}\Big(\frac{t}{2^j r_B}\Big)^{n+1} dt \\ &\leq \ C2^{-j(n+1)}|B|^{-1/2}, \end{aligned}$$

where the last step makes use of the condition $M > n$. Combining estimates (8.25)-(8.26) we have that $\|m\|_{L^2(U_j(B))} \leq C2^{-j(n+1)}|B|^{-1/2}$. Therefore,

$$\Big\| |x-x_B|^n m \Big\|_{L^2(\mathbb{R}^n)} \leq C\sum_{j=0}^{\infty}(2^j r_B)^n 2^{-j(n+1)}|B|^{-1/2} \leq C|B|^{1/2},$$

which, combined with (8.24), shows that $\nabla L^{-1/2} a$ satisfies condition (8.22). Consequently, $\nabla L^{-1/2} a$ is a molecule. The proof of Theorem 8.6 is therefore complete. $\qquad \square$

Remarks. Let $L = -\Delta + V$ be a Schrödinger operator, where $0 \leq V \in L^1_{\text{loc}}(\mathbb{R}^n)$.

(i) The fact that the Riesz transforms $\nabla L^{-1/2}$ associated with L are of weak type $(1,1)$ can be seen as in [**CD1**]. See also [**Si2**].

(ii) From Theorem 8.6, we have

$$H^1_{L,at,M}(\mathbb{R}^n) \subseteq H^1_{L,\text{Riesz}}(\mathbb{R}^n) := \Big\{ f \in L^1(\mathbb{R}^n) : \nabla L^{-1/2} f \in L^1(\mathbb{R}^n) \Big\}.$$

In general, it remains an open problem to determine whether the reverse inclusion $H^1_{L,\text{Riesz}}(\mathbb{R}^n) \subseteq H^1_{L,at,M}(\mathbb{R}^n)$ holds. However, under the stronger assumption of (local) $L^{n/2+\varepsilon}$ integrability of the potential, this question has been resolved in the affirmative in [**DZ1**] and [**DP**].

CHAPTER 9

Further properties of Hardy spaces associated to operators

9.1. The semigroup with the conservation property. Let (X, d, μ) be as in (2.1), and let L be an operator satisfying **(H1)** and **(H2)**. We claim that e^{-tL} maps L^2 functions with compact supports into L^1. Indeed, fix $t > 0$ and $\phi \in L^2(X)$ supported in a ball B. Let $U_j(B)$ be the annuli defined in (2.6). Using the hypothesis **(H2)**, we have that for every $j \geq 0$,

$$(9.1) \qquad \int_{U_j(B)} |e^{-tL}\phi(x)|d\mu(x) \leq V(U_j(B))^{1/2} \exp\left(- \frac{\text{dist}(U_j(B), B)^2}{ct} \right) \|\phi\|_{L^2(B)},$$

so that summing in j gives the claim. Hence, the action of the semigroup on $L^\infty(X)$ can be defined in the $L^2_{\text{loc}}(X)$ sense via duality. In this section, we assume that for all $t > 0$,

$$\textbf{(H4)} \qquad\qquad e^{-tL}1 = 1, \qquad \text{in } L^2_{\text{loc}}(X).$$

That is, for every $\phi \in L^2(X)$ with compact support,

$$\int_X \left(e^{-tL}1\right)(x)\,\phi(x)d\mu(x) : \; = \; \int_X e^{-tL}\phi(x)d\mu(x)$$

$$(9.2) \qquad\qquad\qquad\qquad = \; \int_X \phi(x)d\mu(x)$$

for all $t > 0$. We have the following:

LEMMA 9.1. *Suppose $M \geq 1$. For an operator L satisfying* **(H1)**, **(H2)** *and* **(H4)**, *then for every $(1, 2, M)$-atom a,*

$$\int_X a(x)d\mu(x) = 0.$$

PROOF. It follows from the hypothesis **(H1)** that L is a non-negative self-adjoint operator in $L^2(X)$. Let ϕ in $L^2(X)$ with compact support (so that, by our previous observation, $e^{-tL}\phi \in L^1(X)$). By L^2 functional calculus, we may write $(I + L)^{-1}\phi = \int_0^\infty e^{-t}e^{-tL}\phi dt$. This and (9.2) yield

$$\int_X (I + L)^{-1}\phi(x)d\mu(x) = \int_0^\infty e^{-t} \left[\int_X e^{-tL}\phi(x)d\mu(x) \right] dt$$

$$(9.3) \qquad\qquad\qquad\qquad = \int_X \phi(x)d\mu(x),$$

where our use of Fubini's theorem in the first equality may be justified by the fact that $e^{-tL}\phi \in L^1(X)$.

Suppose now that a is a $(1, 2, M)$-atom associated to a ball B. By definition, there exists a function $b \in \mathcal{D}(L^M)$, such that $a = L^M b$, which satisfies (ii) and (iii) in Definition 2.1. Set $b_1 = L^{M-1}b$, and thus $a = Lb_1$. Note that $b_1 \in L^2(X)$ is supported in B. We apply (9.3) twice to obtain

$$\int_X a(x)d\mu(x) = \int_X (I + L)^{-1}Lb_1(x)d\mu(x)$$

$$= \int_X (I + L)^{-1}(I + L)b_1(x)d\mu(x) - \int_X (I + L)^{-1}b_1(x)d\mu(x)$$

$$= \int_X b_1(x)d\mu(x) - \int_X b_1(x)d\mu(x) = 0.$$

This proves Lemma 9.1. $\qquad\qquad\square$

9.2. Hardy spaces $H_L^p(X)$ for $1 \leq p < \infty$.

9.2.1. *Definition.* In previous sections, we have treated the Hardy space $H_L^1(X)$ (cf. Definition 4.2) associated to a given operator L. The goal of this section is to study the Hardy spaces $H_L^p(X)$ associated to L for all $1 < p < \infty$, by means of quadratic operators and tent spaces. These spaces were previously introduced in [**ADM**] and [**AMR**] under somewhat more specialized circumstances.

Given an operator L satisfying (**H1**)-(**H2**) and $\widetilde{M} \geq 1$, consider the following quadratic operators associated to L

$$(9.4) \qquad S_{h,\widetilde{M}}f(x) := \left(\iint_{\Gamma(x)} |(t^2 L)^{\widetilde{M}} e^{-t^2 L} f(y)|^2 \frac{d\mu(y)}{V(x,t)} \frac{dt}{t} \right)^{1/2}, \qquad x \in X$$

where $f \in L^2(X)$. Also, abbreviate $S_{h,1}f = S_h f$.

For each $\widetilde{M} \geq 1$ and $1 \leq p < \infty$, now define

$$(9.5) \qquad D_{\widetilde{M},p} := \left\{ f \in H^2(X) : S_{h,\widetilde{M}}f \in L^p(X) \right\}, \qquad 1 \leq p < \infty.$$

DEFINITION 9.2. *Suppose $\widetilde{M} \geq 1$. For $1 \leq p < \infty$, the Hardy space $H_{L,\widetilde{M}}^p(X)$ associated to L (assumed to satisfy (**H1**)-(**H2**)) is the completion of the space $D_{\widetilde{M},p}$ in the norm*

$$\|f\|_{H_{L,\widetilde{M}}^p(X)} := \|S_{h,\widetilde{M}}f\|_{L^p(X)}.$$

For every $\widetilde{M} \geq 1$, it follows from (3.14) for an appropriate choice of ψ that $H_{L,\widetilde{M}}^2(X) = H^2(X)$. Moreover, we have that $H_{L,\widetilde{M}}^1(X) = H_L^1(X)$ for all $\widetilde{M} \geq 1$. Indeed, by definition, $H_{L,1}^1(X) = H_{L,S_h}^1(X)$ and a similar argument to that in Propositions 4.4 and 4.13 shows that for each $\widetilde{M} \geq 1$ and $M > n_0/4$, the spaces $H_{L,\widetilde{M}}^1(X)$ and $H_{L,at,M}^1(X)$ coincide. Under an assumption of Gaussian upper bounds for the heat kernel of the operator L, it was proved in [**ADM**] that for every $\widetilde{M} \geq 1$, $H_{L,\widetilde{M}}^p(X) = L^p(X)$ for all $1 < p < \infty$. Note that, in the framework of the present paper, we only assume the Davies-Gaffney estimates on the heat kernel of L, and hence for $1 < p < \infty$, $p \neq 2$, $H_{L,\widetilde{M}}^p(X)$ may or may not coincide with the space $L^p(X)$.

For every $f \in H_{L,\widetilde{M}}^p(X), 1 < p < \infty$, we now consider

$$(9.6) \qquad Q_{t,L,\widetilde{M}}f(x, t) := (t^2 L)^{\widetilde{M}} e^{-t^2 L} f(x), \qquad t > 0, \quad x \in X.$$

Then the operator $Q_{t,L,\widetilde{M}}$ embeds the Hardy space $H_{L,\widetilde{M}}^p(X)$ isometrically into the tent space $T_2^p(X)$ for $1 < p < \infty$. Of importance shall also be another operator acting to the opposite direction. Let $M \geq 1$. Consider the operator $\pi_{L,M} : T_2^2(X) \to H^2(X)$, given by

$$(9.7) \qquad \pi_{L,M}(F)(x) := \int_0^\infty (t^2 L)^M e^{-t^2 L}\big(F(\cdot, t)\big)(x)\frac{dt}{t},$$

where the improper integral converges in L^2. Then the bound

$$(9.8) \qquad \|\pi_{L,M} F\|_{L^2(X)} \leq C_M \|F\|_{T_2^2(X)}, \quad M \geq 1,$$

follows readily by duality and the L^2 quadratic estimate (3.14). By L^2-functional calculus, for every $f \in H^2(X)$ there exists a constant $c_{M,\widetilde{M}}$ such that we have the "Calderón reproducing formula"

$$(9.9) \qquad f(x) = c_{M,\widetilde{M}} \pi_{L,M}\big(Q_{t,L,\widetilde{M}} f\big)(x), \quad M, \widetilde{M} \geq 1,$$

in $L^2(X)$.

Let $T_{2,c}^p(X)$ be the set of all $f \in T_2^p(X)$ with compact support in $X \times (0, \infty)$. It follows from (4.11) that $\pi_{L,M}$ is well defined, and $\pi_{L,M} F \in H^2(X)$ for all $F \in T_{2,c}^p(X)$.

PROPOSITION 9.3. *Assume the operator L satisfies* **(H1)**-**(H2)**. *Let $M \geq 1$. The operator $\pi_{L,M}$, initially defined on $T_{2,c}^p$, extends to a bounded linear operator from*
  (a) $T_2^1(X)$ *into* $H_L^1(X)$, *if $M > n_0/4$;*
  (b) $T_2^p(X)$ *into* $H_{L,\widetilde{M}}^p(X)$ *if $1 < p < 2$, $\widetilde{M} \geq 1$, and $M > n_0/4$;*
  (c) $T_2^p(X)$ *into* $H_{L,\widetilde{M}}^p(X)$ *if $2 < p < \infty$, $\widetilde{M} > n_0/4$, and $M \geq 1$.*

PROOF. Let us first prove (a). For $F \in T_{2,c}^1(X)$, we have that, for some $\delta > 0$ depending on the support of F,

$$
\begin{aligned}
\pi_{L,M}(F)(x) &:= \int_\delta^{1/\delta} (t^2 L)^M e^{-t^2 L}\big(F(\cdot, t)\big)(x)\frac{dt}{t} \\
&= \sum_i \lambda_i \int_\delta^{1/\delta} (t^2 L)^M e^{-t^2 L}\big(A_i(\cdot, t)\big)(x)\frac{dt}{t},
\end{aligned}
$$

where we have used Proposition 4.10 to write $F = \sum \lambda_i A_i$, where the A_i are $T_2^1(X)$ atoms, with $\sum |\lambda_i| \approx \|F\|_{T_2^1}$, and where we have used compactness of the interval of integration in $(0, \infty)$ to justify the interchange of the order of the sum and the integral. It is therefore enough to show that for every $T_2^1(X)$-atom A associated to B and satisfying $\int_{X \times (0,\infty)} |A(x,t)|^2 d\mu(x)dt/t \leq V(B)^{-1}$, we have that, up to a fixed multiplicative constant, $\pi_{L,M}(F)$ is a $(1, 2, M, 1)$-molecule associated to the ball B. To this end, we may write

$$\pi_{L,M}(A) = L^M b,$$

where

$$b = \int_0^\infty t^{2M} e^{-t^2 L}\big(A(\cdot, t)\big)\frac{dt}{t}.$$

Let $\{U_j(B)\}_{j\geq 0}$ be the annuli defined in (2.6). Consider some $g \in L^2(U_j(B))$ such that $\|g\|_{L^2(U_j(B))} = 1$. Then for every $k = 0, 1, \ldots, M$ there holds

$$(9.10) \quad \left| \int_X (r_B^2 L)^k b(x) g(x) d\mu(x) \right| \leq$$

$$\left| \int_{X\times(0,\infty)} t^{2M+2k} L^k e^{-t^2 L}\big(A(\cdot,t)\big)(x) g(x) \frac{d\mu(x) dt}{t} \right|$$

$$\leq \left| \int_{\widehat{B}} A(x,t) t^{2M+2k} L^k e^{-t^2 L} g(x) \frac{d\mu(x) dt}{t} \right|$$

$$\leq r_B^{2M} \|A\|_{T_2^2(X)} \left(\int_{\widehat{B}} \left| (t^2 L)^k e^{-t^2 L} g(x) \right|^2 \frac{d\mu(x) dt}{t} \right)^{1/2}$$

$$\leq C r_B^{2M} V(B)^{-1/2} \|g\|_{L^2(U_j(B))}.$$

Note that the third inequality is obtained by using the fact that A is a T_2^1-atom supported in $\widehat{B}$, hence, $0 < t < r_B$, and that the last inequality follows from (3.14). This gives for every $j = 0, 1, 2$,

$$\|(r_B^2 L)^k b\|_{L^2(U_j(B))} \leq C r_B^{2M} V(B)^{-1/2}.$$

Fix $j \geq 3$, with the goal of estimating the L^2-norm of $(r_B^2 L)^k b$ on $U_j(B)$. Arguing as in (9.10) and invoking condition (**H2**) gives

$$\left| \int_X (r_B^2 L)^k b(x) g(x) d\mu(x) \right|$$

$$\leq r_B^{2M} \|A\|_{T_2^2(X)} \left(\int_{\widehat{B}} \left| (t^2 L)^k e^{-t^2 L} g(x) \right|^2 \frac{d\mu(x) dt}{t} \right)^{1/2}$$

$$\leq C r_B^{2M} V(B)^{-1/2} \left(\int_0^{r_B} \left\| (t^2 L)^k e^{-t^2 L} g \right\|_{L^2(B)}^2 \frac{dt}{t} \right)^{1/2}$$

$$\leq C r_B^{2M} V(B)^{-1/2} \left(\int_0^{r_B} e^{-\frac{\operatorname{dist}(U_j(B),B)^2}{ct^2}} \|g\|_{L^2(U_j(B))}^2 \frac{dt}{t} \right)^{1/2}$$

$$\leq C r_B^{2M} V(B)^{-1/2} \left(\int_0^{r_B} \left(\frac{t}{2^j r_B} \right)^{2(n_0+1)} \frac{dt}{t} \right)^{1/2}$$

$$\leq C r_B^{2M} V(B)^{-1/2} 2^{-j(n_0+1)}$$

$$\leq C 2^{-j} r_B^{2M} V(2^j B)^{-1/2}.$$

Hence, $\pi_{L,M}(F)$ is a constant multiple of a $(1, 2, M, 1)$-molecule associated to the ball B. This proves (a).

We note that (a) (resp. (b) and (c)) is equivalent to the statement that the mapping $Q_{t,L,\widetilde{M}} \circ \pi_{L,M}$ is bounded on $T_2^1(X)$ (resp. $T_2^p(X)$). Of course, a similar statement applies to $T_2^2(X)$, given (9.8) and (3.14). The case $1 < p < 2$ (i.e., conclusion (b)) now follows by the interpolation result for tent spaces (Proposition 4.8). The statement in (c) is a consequence of the tent space duality (Proposition 4.7), along with the observation that, viewed as a bounded mapping on T_2^p, the adjoint of $Q_{t,L,\widetilde{M}} \circ \pi_{L,M}$ is $Q_{t,L,M} \circ \pi_{L,\widetilde{M}}$, as the reader may readily verify. This completes the proof of Proposition 9.3. $\qquad\square$

As a consequence of the previous Proposition, we have the following duality result.

PROPOSITION 9.4. *Suppose that $1 < p < \infty$, and that $\widetilde{M}(p) \geq 1$, $1 < p \leq 2$, and $\widetilde{M}(p) \geq n_0/4$, $2 < p < \infty$. Then the dual of $H^p_{L,\widetilde{M}(p)}(X)$ is $H^{p'}_{L,\widetilde{M}(p')}(X)$, with $1/p + 1/p' = 1$. More precisely, the pairing $\langle f, g \rangle \mapsto \int_X f(x)g(x)d\mu(x)$, realizes $H^{p'}_{L,\widetilde{M}(p')}(X)$ as equivalent to the dual of $H^p_{L,\widetilde{M}(p)}(X)$.*

SKETCH OF PROOF. We follow [**CMS**]. To show that $H^{p'}_{L,\widetilde{M}(p')} \subseteq (H^p_{L,\widetilde{M}(p)})^*$, we take g in the dense class $H^{p'}_{L,\widetilde{M}(p')} \cap H^2$, and $f \in H^p_{L,\widetilde{M}(p)} \cap H^2$, and use the Calderón reproducing formula (9.9), then Cauchy-Schwarz's inequality in t and Hölder's inequality in x to bound $|\langle f, g \rangle|$ by the product of the L^p and $L^{p'}$ norms of the square functions of f and g. We omit the routine details.

Consider now the other direction, i.e. $(H^p_{L,\widetilde{M}(p)})^* \subseteq H^{p'}_{L,\widetilde{M}(p')}$. Suppose that $\Lambda \in (H^p_{L,\widetilde{M}(p)})^*$. We identify $H^p_{L,\widetilde{M}(p)}$ with a subspace of $T^p_2(X)$ via the mapping $f \to (t^2 L)^{\widetilde{M}(p)} e^{-t^2 L} f$. Then by the Hahn-Banach Theorem, we may extend Λ to $\widetilde{\Lambda} \in (T^p_2(X))^* = T^{p'}_2(X)$ so that

$$\widetilde{\Lambda}\left((t^2 L)^{\widetilde{M}(p)} e^{-t^2 L} f \right) = \Lambda(f).$$

Thus, there exists $G \in T^{p'}_2$ such that for $F \in T^p_2$,

$$\widetilde{\Lambda}(F) = \int_0^\infty \int_X G(x,t)F(x,t)d\mu(x)\frac{dt}{t}.$$

In particular, for $F(x,t) := (t^2 L)^{\widetilde{M}(p)} e^{-t^2 L} f$, we have

$$\Lambda(f) = \int_0^\infty \int_X G(x,t)\, (t^2 L)^{\widetilde{M}(p)} e^{-t^2 L} f \, d\mu(x)\frac{dt}{t}$$

$$= \int_X \pi_{L,\widetilde{M}(p)}(G)(x)f(x)d\mu(x) = \int_X g(x)f(x)d\mu(x).$$

But $g \in H^{p'}_{L,\widetilde{M}(p')}$, as desired, by Proposition 9.3. $\square$

Turning to the theory of complex interpolation of Hardy spaces, recall that $[\cdot,\cdot]_\theta$ stands for the complex interpolation bracket.

PROPOSITION 9.5. *Let L be an operator satisfying (**H1**)-(**H2**). Suppose $\widetilde{M} \geq 1$, $1 \leq p_0 < p_1 < \infty$, $0 < \theta < 1$, and $1/p = (1-\theta)/p_0 + \theta/p_1$. Then*

$$[H^{p_0}_{L,\widetilde{M}}(X), H^{p_1}_{L,\widetilde{M}}(X)]_\theta = H^p_{L,\widetilde{M}}(X)$$

(provided in addition that $\widetilde{M} > n_0/4$ if $p_1 > 2$).

PROOF. This follows from the following general principle (see Theorem I.2.4, [**Tr**]): Let X_0, X_1 and Y_0, Y_1 be two interpolation couples such that there exist operators $S \in \mathcal{L}(Y_i, X_i)$ and $Q \in \mathcal{L}(X_i, Y_i)$ with $SQx = x$ for all $x \in X_i$ and $i = 0, 1$. Then $[X_0, X_1]_\theta = S[Y_0, Y_1]_\theta$. Here we take $S = c_{M,\widetilde{M}}\pi_{L,M}$, $M > n_0/4$ and $Q = Q_{t,L,\widetilde{M}}$, in combination with (9.9) and Proposition 9.3. $\square$

Remark: Since for every $\widetilde{M} \geq 1$, we have that $H^1_{L,\widetilde{M}}(X) = H^1_{L,1}(X)$ and $H^2_{L,\widetilde{M}}(X) = H^2(X)$, it follows from Proposition 9.5 that for every $\widetilde{M} \geq 1$, $H^p_{L,\widetilde{M}}(X) = H^p_{L,1}(X)$ for all $1 \leq p \leq 2$, and, by Proposition 9.4, that $H^p_{L,\widetilde{M}}(X) = H^p_{L,M_0}(X)$ for all $2 < p < \infty$ and $M_0 = [\frac{n_0}{4}] + 1$. We are now able to give the following definition of the $H^p_L(X)$ for all $1 \leq p < \infty$ (see also Section 5, [**AMR**]).

DEFINITION 9.6. *Let L be an operator satisfying* (**H1**)-(**H2**).

(i) For each $1 \leq p \leq 2$, the Hardy space $H^p_L(X)$ associated with L is the completion of the space $D_{1,p}$ in the norm

$$\|f\|_{H^p_L(X)} := \|S_h f\|_{L^p(X)}.$$

(ii) For each $2 < p < \infty$, the Hardy space $H^p_L(X)$ associated with L is the completion of the space $D_{M_0,p}$ in the norm

$$\|f\|_{H^p_L(X)} := \|S_{h,M_0} f\|_{L^p(X)}, \qquad M_0 = [\frac{n_0}{4}] + 1.$$

Remark: Recall that $H^2_L(X) = H^2(X) \subset L^2(X)$. On the other hand, it remains an open problem, in this general context, to determine whether $H^p_L(X) \subseteq L^p(X)$ for $1 \leq p < 2$. Of course, $H^2_L(X)$ embeds continuously into $L^2(X)$, and moreover, we know that $H^2_L(X) \cap H^1_L(X)$ embeds continuously into L^1. Indeed, by Theorem 4.1 and its proof (and the definition of $\mathbb{H}^1_{L,at,M}(X)$), we have

$$\|f\|_{L^1(X)} \leq C\|f\|_{H^1_L(X)}, \qquad f \in H^2_L(X) \cap H^1_L(X).$$

Thus, extending by continuity, we may deduce the existence of a continuous map $\mathcal{J} : H^1_L(X) \to L^1(X)$, which equals the identity on $H^2_L(X) \cap H^1_L(X)$, but in general it remains an open problem to determine whether this map $\mathcal{J}$ is 1-1. At present, one can at least say that this embedding is 1-1 in the special case of the Laplace-Beltrami operator on a Riemannian manifold with a doubling measure [**AMc**], and also in general under the stronger pointwise Gaussian heat kernel bound condition (**H3**) (cf. (7.1)). We sketch now an argument to establish this fact in the latter case. Interpolating the inclusion map will then yield more generally that $H^p_L(X) \subseteq L^p(X)$ for $1 \leq p < 2$, in the presence of a pointwise Gaussian heat kernel bound.

Let $f \in H^1_L(X)$. Then there is an atomic decomposition $f = \sum \lambda_i a_i$ converging to f in $H^1_L(X)$, with $\sum |\lambda_i| \approx \|f\|_{H^1_L(X)}$. Moreover, the partial sums $f_N := \sum_{i=1}^N \lambda_i a_i$ belong to $H^2(X) \cap H^1_L(X)$, and by Theorem 5.4 we have

$$\|f_N\|_{L^1(X)} \leq C\|f_N\|_{H^1_L(X)}.$$

Since $f_N \to f$ in $H^1_L(X)$, we may make an extension by continuity to obtain $\mathcal{J}f \in L^1$ such that $\|\mathcal{J}f\|_{L^1(X)} \leq \|f\|_{H^1_L(X)}$, with $f_N \to \mathcal{J}f$ in $L^1(X)$. On the other hand, the atomic sum clearly converges in $L^1(X)$, so that $\mathcal{J}f = \sum_{i=1}^\infty \lambda_i a_i$ in $L^1(X)$. But by a vector-valued version of the weak-type (1,1) estimates of Duong [**DM**], the square function S_h is of weak-type (1,1) (here we are using the pointwise heat kernel bounds), so that $S_h(\mathcal{J}f - f_N) \to 0$ in the weak-L^1 space $L^{1,\infty}(X)$. On the other hand, $S_h(f_N - f) \to 0$ in $L^1(X)$, as this is equivalent to the fact that $f_N \to f$ in $H^1_L(X)$. Thus, if $\mathcal{J}f = 0$ in $L^1(X)$, then $f = 0$ in $H^1_L(X)$, i.e., the embedding map $\mathcal{J}$ is 1-1.

The obstacle to extending this argument to the general case in which pointwise kernel bounds may be lacking is the absence of weak-type (1,1) estimates for the square function.

9.2.2. *An interpolation theorem.* Let L be an operator satisfying **(H1)-(H2)**. We shall now discuss a Marcinkiewicz-type interpolation theorem. Other interpolation theorems for generalized Hardy spaces have been obtained in **[BeZ]** and **[Be]**. In order to state the next result, we first need to recall the concept of weak-type operators. If T is defined on $H_L^p(X)$, for some $p \geq 1$, we say that it is of weak-type (H_L^p, p) provided

$$\mu\{x \in X : |Tf(x)| > \lambda\} \leq C\lambda^{-p}\|f\|_{H_L^p(X)}^p$$

for all $f \in H_L^p(X)$. The best constant C will be referred to as being the weak-type norm of T. We can now state the following

THEOREM 9.7. *Let L be an operator satisfying* **(H1)-(H2)**. *Suppose* $1 \leq p_1 \leq p_2 < \infty$, $p_1 < p_2$, *and let T be a sublinear operator from $H_L^{p_1}(X) + H_L^{p_2}(X)$ into measurable functions on X, which is of weak-type $(H_L^{p_1}, p_1)$ and $(H_L^{p_2}, p_2)$ with weak-type norms C_1 and C_2, respectively. If $p_1 < p < p_2$, then T is bounded from $H_L^p(X)$ into $L^p(X)$ and*

$$(9.11) \qquad \|Tf\|_{L^p(X)} \leq C\|f\|_{H_L^p(X)},$$

where C depends only on C_1, C_2, p_1, p_2, and p.

PROOF. Fix $p \in (p_1, p_2)$. It is enough to establish (9.11) for f in the dense class $H^2(X) \cap H_L^p(X)$. By the remark preceding Definition 9.6, $S_{h,M_0}f \in L^p(X)$ so that

$$F(x,t) := (t^2 L)^{M_0} e^{-t^2 L} f \in T_2^p(X), \quad \text{where } M_0 = \left[\frac{n_0}{4}\right] + 1.$$

Following the proof of Theorem 4' of **[CMS]**, for every $\lambda > 0$ we let $O_\lambda = \{x \in X : \mathcal{A}^3 F(x) > \lambda\}$ (here, the superscript "3" refers to the aperture of the cone defining the square function; see (4.10)), and write $F = F^\lambda + F_\lambda$, where

$$(9.12) \qquad F^\lambda = \chi_{\widehat{O}_\lambda} F \quad \text{and} \quad F_\lambda = \chi_{X \times (0,\infty) \setminus \widehat{O}_\lambda} F$$

(recall the tent spaces defined in (4.9)). Observe that $\mathcal{A}F^\lambda(x) \leq \mathcal{A}F(x) \leq \mathcal{A}^3 F(x)$ for all $x \in X$ and $\mathcal{A}F^\lambda$ is supported only in O_λ, i.e., where $\mathcal{A}^3 F(x) > \lambda$. We also have $\mathcal{A}F_\lambda(x) \leq \mathcal{A}F(x) \leq \mathcal{A}^3 F(x)$, $x \in X$, so that $\mathcal{A}F_\lambda(x) \leq \lambda$ for $x \in (O_\lambda)^c$. We now claim that the same bound holds for $\mathcal{A}F_\lambda$ in O_λ, hence that

$$(9.13) \qquad \mathcal{A}F_\lambda(x) \leq \lambda, \quad x \in X.$$

Indeed, a simple geometric argument shows that if $x \in O_\lambda$, and $\bar{x}$ is a point in $(O_\lambda)^c$ of minimum distance from x, then $\Gamma(x) \cap (\widehat{O}_\lambda)^c \subset \Gamma^3(\bar{x})$. Thus $\mathcal{A}F_\lambda(x) \leq \mathcal{A}^3 F(\bar{x}) \leq \lambda$. Let $\pi_{L,M}$ be as in (4.16) and c_{M,M_0} is the constant in (9.9). Fix $M > n_0/4$. For every $\lambda > 0$ define

$$f^\lambda := c_{M,M_0} \pi_{L,M}(F^\lambda) \quad \text{and} \quad f_\lambda := c_{M,M_0} \pi_{L,M}(F_\lambda).$$

It follows from (9.9) and the definition of F that $f = f^\lambda + f_\lambda$. Let us look separately at the two terms of this decomposition.

(a) $f^\lambda \in H_L^{p_1}(X)$. Indeed, it follows from the properties of $\mathcal{A}F^\lambda$ and Lemma 4.6 that

$$\|\mathcal{A}F^\lambda\|_{L^{p_1}(X)}^{p_1} \leq \lambda^{p_1-p}\|\mathcal{A}^3 F\|_{L^p(X)}^p \leq C\lambda^{p_1-p}\|\mathcal{A}F\|_{L^p(X)}^p \leq C\lambda^{p_1-p}\|f\|_{H_L^p(X)}^p$$

which gives $F^\lambda \in T_2^{p_1}(X)$. This, together with Proposition 9.3, implies that

$$\|f^\lambda\|_{H_L^{p_1}(X)} \leq C\lambda^{p_1-p}\|f\|_{H_L^p(X)}^p.$$

(b) $f_\lambda \in H_L^{p_2}(X)$. This is seen by first using (9.13) to deduce that

$$\|\mathcal{A}F_\lambda\|_{L^{p_2}(X)}^{p_2} \leq \lambda^{p_2-p}\|\mathcal{A}F\|_{L^p(X)}^p = C\lambda^{p_2-p}\|f\|_{H_L^p(X)}^p,$$

whence by Proposition 9.3 it follows that

$$\|f_\lambda\|_{H_L^{p_1}(X)} \leq C\lambda^{p_2-p}\|f\|_{H_L^p(X)}^p.$$

Since T is sublinear, we have that $|Tf| \leq |Tf^\lambda| + |Tf_\lambda|$ for every $\lambda > 0$. Then the weak-type hypothesis implies

$$(9.14) \quad p^{-1}\|Tf\|_{L^p(X)}^p = \int_0^\infty \lambda^{p-1}\mu\{x \in X : Tf(x) > \lambda\}\, d\lambda$$

$$\leq \int_0^\infty \lambda^{p-1}\mu\{x \in X : Tf^\lambda(x) > \lambda/2\}\, d\lambda$$

$$+ \int_0^\infty \lambda^{p-1}\mu\{x \in X : Tf_\lambda(x) > \lambda/2\}\, d\lambda$$

$$\leq C \int_0^\infty \lambda^{p-1}\Big(\frac{2C_1\|f^\lambda\|_{H_L^{p_1}(X)}}{\lambda}\Big)^{p_1}\, d\lambda$$

$$+ C \int_0^\infty \lambda^{p-1}\Big(\frac{2C_2\|f_\lambda\|_{H_L^{p_2}(X)}}{\lambda}\Big)^{p_2}\, d\lambda$$

$$\leq CC_1^{p_1} \int_0^\infty \lambda^{p-1-p_1}\|f^\lambda\|_{H_L^{p_1}(X)}^{p_1}\, d\lambda + CC_2^{p_2} \int_0^\infty \lambda^{p-1-p_2}\|f_\lambda\|_{H_L^{p_2}(X)}^{p_2}\, d\lambda.$$

To estimate the first integral in the last line of (9.14), we make use of Proposition 9.3 and the fact that $f^\lambda = c_{M,M_0}\pi_{L,M}(F^\lambda)$ in order to majorize it by

$$(9.15) \quad \int_0^\infty \lambda^{p-1-p_1}\|\pi_{L,M}(F^\lambda)\|_{H_L^{p_1}(X)}^{p_1}\, d\lambda \leq C \int_0^\infty \lambda^{p-1-p_1}\|F^\lambda\|_{T_2^{p_1}(X)}^{p_1}\, d\lambda$$

$$= C \int_0^\infty \lambda^{p-1-p_1}\Big\{\int_{\{x:\, \mathcal{A}^3 F(x) > \lambda\}} |\mathcal{A}F^\lambda(x)|^{p_1}\, d\mu(x)\Big\}\, d\lambda,$$

since $\mathcal{A}F^\lambda$ is supported in $O_\lambda = \{x \in X : \mathcal{A}^3 F(x) > \lambda\}$. Using the fact that $\mathcal{A}F^\lambda(x) \leq \mathcal{A}^3 F(x)$ and Fubini's theorem, the last integral above is further bounded by

$$\int_X |\mathcal{A}^3 F(x)|^{p_1}\Big\{\int_0^{\mathcal{A}^3 F(x)} \lambda^{p-1-p_1}\, d\lambda\Big\}\, d\mu(x)$$

$$(9.16) \quad \leq \frac{1}{p-p_1}\|\mathcal{A}^3 F\|_{L^p(X)}^p \leq \frac{C}{p-p_1}\|\mathcal{A}F\|_{L^p(X)}^p = \frac{C}{p-p_1}\|f\|_{H_L^p(X)}^p,$$

where the second inequality uses Lemma 4.6.

In order to estimate the second integral in the last line of (9.14), we make use of Proposition 9.3 and the fact that $f_\lambda = c_{M,M_0} \pi_{L,M}(F_\lambda)$ to write

$$\int_0^\infty \lambda^{p-1-p_2} \|f_\lambda\|_{H_L^{p_2}(X)}^{p_2} \, d\lambda \;\leq\; C \int_0^\infty \lambda^{p-1-p_2} \|\pi_{L,M}(F_\lambda)\|_{H_L^{p_2}(X)}^{p_2} d\lambda$$

$$\leq\; C \int_0^\infty \lambda^{p-1-p_2} \|\mathcal{A}F_\lambda\|_{L^{p_2}(X)}^{p_2} d\lambda.$$

As observed before, we have that $\mathcal{A}F_\lambda(x) \leq \lambda$ for all $x \in X$. Also, trivially from (9.12), $\mathcal{A}F_\lambda(x) \leq \mathcal{A}^3 F(x)$ for $x \in X$. Using these observations, (9.16) and Fubini's theorem once again, we can dominate the last integral above by

$$\int_0^\infty \lambda^{p-1-p_1} \left\{ \int_{\{x : \mathcal{A}^3 F(x) > \lambda\}} |\mathcal{A}^3 F(x)|^{p_1} d\mu(x) \right\} d\lambda$$

$$+ \int_0^\infty \lambda^{p-1-p_2} \left\{ \int_{\{x : \mathcal{A}^3 F(x) \leq \lambda\}} |\mathcal{A}^3 F(x)|^{p_2} d\mu(x) \right\} d\lambda$$

$$\leq\; \frac{C}{p-p_1} \|f\|_{H_L^p(X)}^p + \int_X |\mathcal{A}^3 F(x)|^{p_2} \left\{ \int_{\mathcal{A}^3 F(x)}^\infty \lambda^{p-1-p_2} d\lambda \right\} d\mu(x)$$

$$\leq\; \frac{C}{p-p_1} \|f\|_{H_L^p(X)}^p + \frac{C}{p_2-p} \|\mathcal{A}^3 F\|_{L^p(X)}^p$$

$$\leq\; \left(\frac{C}{p-p_1} + \frac{C}{p_2-p} \right) \|f\|_{H_L^p(X)}^p.$$

Collecting all these estimates we obtain the desired inequality $\|Tf\|_{L^p(X)}^p \leq C\|f\|_{H_L^p(X)}^p$, where the constant C depends only on C_1, C_2, p_1, p_2 and p. $\qquad\square$

Remarks. Assume that L is an operator satisfying **(H1)** and **(H2)**. As consequences of Theorem 9.7, we have the following results of intrinsic importance.

   (i) Based on the computations from § 4.5, one can see that the operator $\mathcal{N}_h$ from (2.11) maps $L^p(X)$ into $L^p(X)$ for $p > 2$ and $L^2(X)$ into weak-$L^2(X)$. Furthermore, by (4.26), it maps $H_L^1(X)$ into $L^1(X)$. Thus, by Theorem 9.7, $\mathcal{N}_h$ also maps $H_L^p(X)$ into $L^p(X)$ for $p \in (1, 2)$.

   (ii) Given a function $f \in L^2(X)$, consider the following vertical version of the square function associated with the heat semigroup generated by L:

$$(9.17) \qquad g_h f(x) := \left(\int_0^\infty |t^2 L e^{-t^2 L} f(x)|^2 \frac{dt}{t} \right)^{1/2}, \qquad x \in X.$$

It follows from (3.14) that the operator g_h is bounded on $L^2(X)$. Arguing as in Proposition 4.4 one can prove that the operator g_h is bounded from $H_L^1(X)$ into $L^1(X)$ and, hence, maps $H_L^p(X)$ into $L^p(X)$ whenever $1 \leq p \leq 2$.

Bibliography

[Au]   P. Auscher, On necessary and sufficient conditions for L^p-estimates of Riesz transforms associated to elliptic operators on $\mathbb{R}^n$ and related estimates, *Memoirs of the Amer. Math. Soc.* 186, no. 871 (2007). MR2292385 (2007k:42025)

[AB]   P. Auscher and B. Ben Ali, Maximal inequalities and Riesz transform estimates on L^p spaces for Schrödinger operators with non-negative potentials, *Annales de l'Institut Fourier*, **57** (2007), 1975-2013. MR2377893 (2009h:35067)

[ADM]   P. Auscher, X.T. Duong and A. McIntosh, Boundedness of Banach space valued singular integral operators and Hardy spaces. Unpublished preprint (2005).

[AHLMT]   P. Auscher, S. Hofmann, M. Lacey, A. McIntosh and Ph. Tchamitchian, The solution of the Kato's square root problem for second elliptic operators on $\mathbb{R}^n$, *Ann. of Math.*, **156** (2002), 633-654. MR1933726 (2004c:47096c)

[AMc]   P. Auscher and A. McIntosh, personal communication.

[AMR]   P. Auscher, A. McIntosh and E. Russ, Hardy spaces of differential forms on Riemannian manifolds, *J. Geom. Anal.*, **18** (2008), 192-248. MR2365673 (2009d:42053)

[AR]   P. Auscher and E. Russ, Hardy spaces and divergence operators on strongly Lipschitz domains of $\mathbb{R}^n$, *J. Funct. Anal.*, **201** (2003), 148-184. MR1986158 (2004c:42049)

[BL]   J. Bergh and J. Löfström, *Interpolation spaces*, Springer-Verlag, New York/Berlin, 1976. MR0482275 (58:2349)

[Be]   F. Bernicot, Use of abstract Hardy spaces, real interpolation and application to bilinear operators, preprint.

[BeZ]   F. Bernicot and J. Zhao, New abstract Hardy spaces, *J. Funct. Anal.*, **255** (2008), 1761-1796. MR2442082 (2009k:46043)

[BK1]   S. Blunck, P. Kunstmann, Weak-type (p,p) estimates for Riesz transforms, *Math. Z.*, 247 (2004), no. 1, 137–148. MR2054523 (2005f:35071)

[BK2]   S. Blunck, P. Kunstmann, Calderón-Zygmund theory for non-integral operators and the H^∞ functional calculus, *Rev. Mat. Iberoamericana*, **19** (2003), 919–942. MR2053568 (2005f:42033)

[B]   M. Bownik, Boundedness of operators on Hardy spaces via atomic decomposition, *Proc. Amer. Math. Soc.*, **133** (2005), 3535-3542. MR2163588 (2006d:42028)

[CT]   A.P. Calderón and A. Torchinsky, Parabolic maximal function associated with a distribution, *Adv. Math.*, **16** (1975), 1-64. MR0417687 (54:5736)

[CarMSp]   Carbonaro, A., Metafune, G., and Spina, C., Parabolic Schrödinger operators, *J. Math. Anal. Appl.* **343**, (2008), 965-974. MR2417116 (2009f:35119)

[CCT]   J. Cheeger, M. Gromov and M. Taylor, Finite propagation speed, kernel estimates for functions of the Laplacian and the geometry of complete Riemannian manifolds, *J. Differential Geom.*, **17** (1982), 15-53. MR658471 (84b:58109)

[Ch]   M. Christ, A $T(b)$ theorem with remarks on analytic capacity and the Cauchy integral, *Colloq. Math.*, **LX/LXI** (1990), 601-628. MR1096400 (92k:42020)

[C]   R. Coifman, A real variable characterization of H^p, *Studia Math.*, **51**(1974), 269-274. MR0358318 (50:10784)

[CMS]   R.R. Coifman, Y. Meyer and E.M. Stein, Some new functions and their applications to harmonic analysis, *J. Funct. Analysis,* **62**(1985), 304-315. MR791851 (86i:46029)

[CW1]   R. Coifman and G. Weiss, *Analyse harmonique non-commutative sur certains espaces homogènes*, Lecture Notes in Mathematics **242**. Springer, Berlin-New York, 1971. MR0499948 (58:17690)

[CW2]   R. Coifman and G. Weiss, Extensions of Hardy spaces and their use in analysis, *Bull. Amer. Math. Soc.*, **83** (1977), 569–645. MR0447954 (56:6264)

[CD1] T. Coulhon and X.T. Duong, Riesz transforms for $1 \le p \le 2$, *Trans. Amer. Math. Soc.*, **351** (1999), 1151-1169. MR1458299 (99e:58174)

[CD2] T. Coulhon and X.T. Duong, Maximal regularity and kernel bounds: observations on a theorem by Hieber and Prüss, *Adv. Differential Equations*, **5** (2000), 343-368. MR1734546 (2001d:34087)

[CS] T. Coulhon and A. Sikora, Gaussian heat kernel upper bounds via Phragmén-Lindelöf theorem, *Proc. Lond. Math.*, **96** (2008), 507-544. MR2396848

[CYZ] D.-C. Chang, D. Yang and Y. Zhou, Boundedness of sublinear operators on product Hardy spaces and its application, *J. Math. Soc. Japan* **62** (2010), 321-353. MR2648225

[Da1] E.B. Davies, *Heat kernels and spectral theory*, Cambridge Univ. Press, 1989. MR990239 (90e:35123)

[Da2] E.B. Davies, Heat kernel bounds, conservation of probability and the Feller property, *J. Anal. Math.*, **58** (1992), 99-119. MR1226938 (94e:58136)

[Da3] E.B. Davies, Non-Gaussian aspects of heat kernel behaviour, *J. London Math. Soc.*, **55** (1997), 105-125. MR1423289 (97i:58169)

[DM] X.T. Duong and A. McIntosh, Singular integral operators with non-smooth kernels on irregular domains, *Rev. Mat. Iberoamericana*, **15** (1999), 233-265. MR1715407 (2001e:42017a)

[DOY] X.T. Duong, E.M. Ouhabaz and L.X. Yan, Endpoint estimates for Riesz transforms of magnetic Schrödinger operators, *Ark. Mat.*, **44** (2006), 261-275. MR2292721 (2008a:35224)

[DR] X.T. Duong and D.W. Robinson, Semigroup kernels, Poisson bounds, and holomorphic functional calculus, *J. Funct. Anal.*, **142** (1996), 89–128. MR1419418 (97j:47056)

[DY1] X.T. Duong and L.X. Yan, New function spaces of BMO type, the John-Nirenberg inequality, interpolation and applications, *Comm. Pure Appl. Math.* **58** (2005), 1375-1420. MR2162784 (2006i:26012)

[DY2] X.T. Duong and L.X. Yan, Duality of Hardy and BMO spaces associated with operators with heat kernel bounds, *J. Amer. Math. Soc.*, **18**(2005), 943-973. MR2163867 (2006d:42037)

[DGMTZ] J. Dziubański, G. Garrigós, T. Martínez, J. Torrea and J. Zienkiewicz, BMO spaces related to Schrödinger operators with potentials satisfying a reverse Hölder inequality, *Math. Z.*, **249** (2005), 329-356. MR2115447 (2005k:35064)

[DP] J. Dziubański and M. Preisner, Riesz transform characterization of Hardy spaces associated with Schrödinger operators with compactly supported potentials, arXiv:0910.1017, (2009).

[DZ1] J. Dziubański and J. Zienkiewicz, Hardy space H^1 associated to Schrödinger operators with potential satisfying reverse Hölder inequality, *Rev. Mat. Iberoamericana* **15** (1999), 279-296. MR1715409 (2000j:47085)

[DZ2] J. Dziubański and J. Zienkiewicz, Hardy space H^1 for Schrödinger operators with certain potentials, *Studia Math.* **164** (2004), 39-53. MR2079769 (2005f:47097)

[FS] C. Fefferman and E.M. Stein, H^p spaces of several variables, *Acta Math.*, **129** (1972), 137–195. MR0447953 (56:6263)

[FOS] G.B. Folland and E.M. Stein, *Hardy spaces on homogeneous groups*, Mathematical Notes, no. 28, Princeton Univ. Press, Princeton, NJ, 1982. MR657581 (84h:43027)

[Ga] M.P. Gaffney, The conservation property of the heat equation on Riemannian manifolds, *Comm. Pure Appl. Math.* **12** (1959), 1-11. MR0102097 (21:892)

[G] D. Goldberg, A local version of real Hardy spaces, *Duke Math. J.*, **46** (1979), 27-42. MR523600 (80h:46052)

[Gr] A. Grigor'yan, Estimates of heat kernels on Riemannian manifolds, in *Spectral theory and geometry (Edinburgh, 1998)*, London Math. Soc., Lecture Note Ser., **273**, 140-225. Cambridge Univ. Press, Cambridge, 1999. MR1736868 (2001b:58040)

[HZ] Y. Han and K. Zhao, Boundedness of operators on Hardy spaces, *Taiwanese J. Math.*, **14** (2010), 319–327. MR2655771

[HLZ] Y. Han, G. Lu and K. Zhao, Discrete Calderón's identity, atomic decomposition and boundedness criterion of operators on multiparameter Hardy spaces, J. Geom. Anal. **20** (2010), 670–689. MR2610894

[HM] S. Hofmann and S. Mayboroda, Hardy and BMO spaces associated to divergence form elliptic operators, *Math. Ann.*, **344** (2009), 37-116. MR2481054 (2009m:42038)

[HM2]   S. Hofmann and S. Mayboroda, ibid., Correction, to appear.

[HMa]   S. Hofmann and J.M. Martell, L^p bounds for Riesz transforms and square roots associated to second order elliptic operators, *Publ. Mat.*, **47** (2003), 497-515. MR2006497 (2004i:35067)

[JN]   F. John, L. Nirenberg, On functions of bounded mean oscillation, *Comm. Pure Appl. Math.* **14** (1961), 415–426. MR0131498 (24:A1348)

[JY]   R. Jiang and D. Yang, New Orlicz-Hardy spaces associated with divergence form elliptic operators, *J. Funct. Anal.* **258** (2010), 1167–1224. MR2565837

[K]   T. Kato, *Perturbation theory for linear operators,* second edition, Springer Verlag, 1976. MR0407617 (53:11389)

[KP]   C.E. Kenig and J. Pipher, The Neumann problem for elliptic equations with nonsmooth coefficients, *Invent. Math.*, **113** (1993), 447-509. MR1231834 (95b:35046)

[L]   R.H. Latter, A decomposition of $H^p(\mathbb{R}^n)$ in terms of atoms, *Studia Math.*, **62** (1977), 92-102. MR0482111 (58:2198)

[LM]   Z.J. Lou and A. McIntosh, Hardy space of exact forms on $\mathbb{R}^N$, *Trans. Amer. Math. Soc.* **357** (2005), 1469–1496. MR2115373 (2006i:42028)

[Mc]   A. McIntosh, Operators which have an H_∞ functional calculus, *Miniconference on operator theory and partial differential equations (North Ryde, 1986)*, 210-231, Proceedings of the Centre for Mathematical Analysis, Australian National University, **14**. Australian National University, Canberra, 1986. MR912940 (88k:47019)

[MS]   R.A Macias and C. Segovia, A decomposition into atoms of distributions on spaces of homogeneous type, *Adv. in Math.*, **33** (1979), 271–309. MR546296 (81c:32017b)

[M]   J.-M. Martell, Sharp maximal functions associated with approximations of the identity in spaces of homogeneous type and applications, *Studia Math.*, **161**(2004), 113-145. MR2033231 (2005b:42016)

[MSV]   S. Meda, P. Sjögren and M. Vallarino, On the H^1-L^1 boundedness of operators, *Proc. Amer. Math. Soc.*, **136** (2008), 2921-2931. MR2399059 (2009b:42025)

[Ou]   E.M. Ouhabaz, *Analysis of heat equations on domains,* London Math. Soc. Monographs, Vol. 31, Princeton Univ. Press, 2005. MR2124040 (2005m:35001)

[Pe]   G. Pedersen, *Analysis now,* Graduate Texts in Mathematics, **118**, Springer-Verlag, New York, 1989. MR971256 (90f:46001)

[RV]   F. Ricci and J. Verdera, Duality in spaces of finite linear combinations of atoms, preprint.

[Ru]   E. Russ, The atomic decomposition for tent spaces on spaces of homogeneous type, *Asymptotic Geometric Analysis, Harmonic Analysis, and Related Topics*, 125-135, Proceedings of the Centre for Mathematical Analysis, Australian National University, **42**. Australian National University, Canberra, 2007. MR2328517 (2008m:46066)

[Sh]   Z.W. Shen, L^p estimates for Schrödinger operators with certain potentials, *Ann. Inst. Fourier (Grenoble)*, **45** (1995), 513-546. MR1343560 (96h:35037)

[Si1]   A. Sikora, On-diagonal estimates on Schrödinger semigroup kernels and reduced heat kernels, *Comm. Math. Phys.*, **188** (1997), 233-249. MR1471338 (98k:58213)

[Si2]   A. Sikora, Riesz transform, Gaussian bounds and the method of wave equation, *Math. Z.*, **247** (2004), 643-662. MR2114433 (2005j:58034)

[Sim]   B. Simon, Maximal and minimal Schrödinger forms, *J. Op. Theory*, **1** (1979), 37-47. MR526289 (81m:35104)

[St1]   E.M. Stein, *Singular integral and differentiability properties of functions,* Princeton Mathematics Series, no. 30, Princeton Univ. Press, 1970. MR0290095 (44:7280)

[St2]   E.M. Stein, *Harmonic analysis: Real variable methods, orthogonality and oscillatory integrals,* Princeton Univ. Press, Princeton, NJ, 1993. MR1232192 (95c:42002)

[St3]   E.M. Stein, Maximal functions II: homogeneous curves, *Proc. Nat. Acad. Sci USA* **73** (1976), 2176-2177. MR0420117 (54:8133b)

[Stu]   K.-Th. Sturm, Analysis on local Dirichlet spaces. II. Upper Gaussian estimates for the fundamental solutions of parabolic equations, *Osaka J. Math.*, **32** (1995), 275-312. MR1355744 (97b:35003)

[SW]   E.M. Stein, G. Weiss, On the theory of harmonic functions of several variables. I. The theory of H^p-spaces, *Acta Math.*, **103** (1960), 25–62. MR0121579 (22:12315)

[TW]   M. Taibleson, and G. Weiss, The molecular characterization of certain Hardy spaces.
          Representation theorems for Hardy spaces, *Astérisque*, **77** (1990), 67-149. MR604370
          (83g:42012)

[T]   M. Taylor, L^p estimates on functions of the Laplace operator, *Duke Math. J.*, **58**(1989),
          773-793. MR1016445 (91d:58253)

[Tr]   H. Triebel, *Interpolation theory, function spaces, differential operators,* North Holland,
          1978. MR503903 (80i:46032b)

[Wi]   J. M. Wilson, On the atomic decomposition for Hardy spaces, *Pacific J. Math.*, **116**
          (1985), no. 1, 201–207. MR769832 (86h:42039)

[YZ]   D.C. Yang and Y. Zhou, A bounded criterion via atoms for linear operators in Hardy
          spaces, *Constr. Approx.*, **29** (2009), 207-218. MR2481589 (2010e:42021)

[Yo]   K. Yosida, *Functional Analysis* (Fifth edition), Springer-Verlag, Berlin, 1978.
          MR0500055 (58:17765)

Editorial Information

To be published in the *Memoirs*, a paper must be correct, new, nontrivial, and significant. Further, it must be well written and of interest to a substantial number of mathematicians. Piecemeal results, such as an inconclusive step toward an unproved major theorem or a minor variation on a known result, are in general not acceptable for publication.

Papers appearing in *Memoirs* are generally at least 80 and not more than 200 published pages in length. Papers less than 80 or more than 200 published pages require the approval of the Managing Editor of the Transactions/Memoirs Editorial Board. Published pages are the same size as those generated in the style files provided for $\mathcal{AMS}$-LaTeX or $\mathcal{AMS}$-TeX.

Information on the backlog for this journal can be found on the AMS website starting from `http://www.ams.org/memo`.

A Consent to Publish and Copyright Agreement is required before a paper will be published in the *Memoirs*. After a paper is accepted for publication, the Providence office will send a Consent to Publish and Copyright Agreement to all authors of the paper. By submitting a paper to the *Memoirs*, authors certify that the results have not been submitted to nor are they under consideration for publication by another journal, conference proceedings, or similar publication.

Information for Authors

Memoirs is an author-prepared publication. Once formatted for print and on-line publication, articles will be published as is with the addition of AMS-prepared frontmatter and backmatter. Articles are not copyedited; however, confirmation copy will be sent to the authors.

Initial submission. The AMS uses Centralized Manuscript Processing for initial submissions. Authors should submit a PDF file using the Initial Manuscript Submission form found at `www.ams.org/submission/memo`, or send one copy of the manuscript to the following address: Centralized Manuscript Processing, MEMOIRS OF THE AMS, 201 Charles Street, Providence, RI 02904-2294 USA. If a paper copy is being forwarded to the AMS, indicate that it is for *Memoirs* and include the name of the corresponding author, contact information such as email address or mailing address, and the name of an appropriate Editor to review the paper (see the list of Editors below).

The paper must contain a *descriptive title* and an *abstract* that summarizes the article in language suitable for workers in the general field (algebra, analysis, etc.). The *descriptive title* should be short, but informative; useless or vague phrases such as "some remarks about" or "concerning" should be avoided. The *abstract* should be at least one complete sentence, and at most 300 words. Included with the footnotes to the paper should be the 2010 *Mathematics Subject Classification* representing the primary and secondary subjects of the article. The classifications are accessible from `www.ams.org/msc/`. The Mathematics Subject Classification footnote may be followed by a list of *key words and phrases* describing the subject matter of the article and taken from it. Journal abbreviations used in bibliographies are listed in the latest *Mathematical Reviews* annual index. The series abbreviations are also accessible from `www.ams.org/msnhtml/serials.pdf`. To help in preparing and verifying references, the AMS offers MR Lookup, a Reference Tool for Linking, at `www.ams.org/mrlookup/`.

Electronically prepared manuscripts. The AMS encourages electronically prepared manuscripts, with a strong preference for $\mathcal{AMS}$-LaTeX. To this end, the Society has prepared $\mathcal{AMS}$-LaTeX author packages for each AMS publication. Author packages include instructions for preparing electronic manuscripts, samples, and a style file that generates the particular design specifications of that publication series. Though $\mathcal{AMS}$-LaTeX is the highly preferred format of TeX, author packages are also available in $\mathcal{AMS}$-TeX.

Authors may retrieve an author package for *Memoirs of the AMS* from `www.ams.org/journals/memo/memoauthorpac.html` or via FTP to `ftp.ams.org` (login as `anonymous`, enter your complete email address as password, and type `cd pub/author-info`). The

AMS Author Handbook and the *Instruction Manual* are available in PDF format from the author package link. The author package can also be obtained free of charge by sending email to `tech-support@ams.org` or from the Publication Division, American Mathematical Society, 201 Charles St., Providence, RI 02904-2294, USA. When requesting an author package, please specify $\mathcal{AMS}$-LaTeX or $\mathcal{AMS}$-TeX and the publication in which your paper will appear. Please be sure to include your complete mailing address.

After acceptance. The source files for the final version of the electronic manuscript should be sent to the Providence office immediately after the paper has been accepted for publication. The author should also submit a PDF of the final version of the paper to the editor, who will forward a copy to the Providence office.

Accepted electronically prepared files can be submitted via the web at `www.ams.org/submit-book-journal/`, sent via FTP, or sent on CD to the Electronic Prepress Department, American Mathematical Society, 201 Charles Street, Providence, RI 02904-2294 USA. TeX source files and graphic files can be transferred over the Internet by FTP to the Internet node `ftp.ams.org` (130.44.1.100). When sending a manuscript electronically via CD, please be sure to include a message indicating that the paper is for the *Memoirs*.

Electronic graphics. Comprehensive instructions on preparing graphics are available at `www.ams.org/authors/journals.html`. A few of the major requirements are given here.

Submit files for graphics as EPS (Encapsulated PostScript) files. This includes graphics originated via a graphics application as well as scanned photographs or other computer-generated images. If this is not possible, TIFF files are acceptable as long as they can be opened in Adobe Photoshop or Illustrator.

Authors using graphics packages for the creation of electronic art should also avoid the use of any lines thinner than 0.5 points in width. Many graphics packages allow the user to specify a "hairline" for a very thin line. Hairlines often look acceptable when proofed on a typical laser printer. However, when produced on a high-resolution laser imagesetter, hairlines become nearly invisible and will be lost entirely in the final printing process.

Screens should be set to values between 15% and 85%. Screens which fall outside of this range are too light or too dark to print correctly. Variations of screens within a graphic should be no less than 10%.

Inquiries. Any inquiries concerning a paper that has been accepted for publication should be sent to `memo-query@ams.org` or directly to the Electronic Prepress Department, American Mathematical Society, 201 Charles St., Providence, RI 02904-2294 USA.

Titles in This Series

TITLES IN THIS SERIES

For a complete list of titles in this series, visit the
AMS Bookstore at **www.ams.org/bookstore/**.